MW00616530

IMPERFECT ENVIRONMENTALIST

IMPERFECT ENVIRONMENTALIST

How to Reduce Waste and Create Change for a Better Planet

SHEILA M. MOROVATI

ROWMAN & LITTLEFIELD
Lanham • Boulder • New York • London

Published by Rowman & Littlefield
An imprint of The Rowman & Littlefield Publishing Group, Inc.
4501 Forbes Boulevard, Suite 200, Lanham, Maryland 20706
www.rowman.com

86-90 Paul Street, London EC2A 4NE

British Library Cataloguing in Publication Information Available

Library of Congress Cataloging-in-Publication Data

Names: Morovati, Sheila, 1978- author.
Title: Imperfect environmentalist : how to reduce waste and create change for a better planet / Sheila Morovati.
Description: Lanham : Rowman & Littlefield, [2024] | Includes bibliographical references and index.
Identifiers: LCCN 2023035742 (print) | LCCN 2023035743 (ebook) | ISBN 9781538179109 (cloth) | ISBN 9781538179116 (ebook)
Subjects: LCSH: Waste minimization—United States—Citizen participation. | Pollution prevention—United States—Citizen participation. | Sustainable living—United States. | Environmentalism—United States. | Morovati, Sheila, 1978- | Environmentalists—United States—Biography.
Classification: LCC TD793.9 .M68 2024 (print) | LCC TD793.9 (ebook) | DDC 640.28/6—dc23/eng/20231120
LC record available at https://lccn.loc.gov/2023035742
LC ebook record available at https://lccn.loc.gov/2023035743

∞™ The paper used in this publication meets the minimum requirements of American National Standard for Information Sciences—Permanence of Paper for Printed Library Materials, ANSI/NISO Z39.48-1992.

I dedicate this book to my two children, Sofia and Leo,
and to all the future generations, who deserve a habitable planet.

CONTENTS

FOREWORD

Sheila Morovati is bad news for cynics.

We all know that the planet is in rough shape. That the system is rigged. That Big Oil and Big Plastic have the politicians in their pockets. That the powers of inertia and greed are insurmountable. That there's nothing we, as individuals, can do but read the horrifying headlines and wish someone in power would *do* something.

That, at least, is how I used to think. I was as cynical as they come. "Replace your light bulbs with LEDs?" Really? *That's* the big contribution a single person can make to saving the earth?

And then Sheila came along.

She is not a scientist. She is not a billionaire. She is not a politician (although maybe she should be).

She is a *single person* who decided to see how far she could go in fixing them. An improbable, crazy, maybe even naive quest, right?

It started with the crayons. Restaurants give a fresh pack to every little kid, so they can draw while they wait for the food to arrive. So cute!

What's not so cute is that crayons are plastic—and the restaurants throw them away once the little kid leaves. 150 million crayons a year go into the dump. It's a national single-use art class.

It occurred to Sheila that the restaurants have a crayon-disposal problem, but LA schools have a not-*enough* crayons problem. She ingeniously engineered a pipeline, whereby the used-once crayons go to kids who really need them. And presto: Millions of pieces of plastic are now diverted from the landfill and delivered with joy to kids who want to draw. A change in the system—produced by *one individual.*

To me, that's news. It's surprising, it's inspiring, and it's a blow to cynicism. Somebody should put that woman on TV!

And so I did, in one of my *CBS Sunday Morning* stories.

But Sheila's crayon thing was only the tip of the landfill iceberg. It was her first foray into one-person activism, but it would not be her last.

Next came her efforts to ban plastic straws (which are, of course, not recyclable) in her town. Spoiler: She succeeded. It was the world's first plastic-straw ban.

Next was her campaign to fix the filthy, broken, abandoned drinking fountains in LA schools. Then came the one where she got companies to stop shipping their already-padded product boxes inside *outer* padded boxes. Then the one where she persuaded movie studios to set a better model for the public—by featuring reusables on camera instead of disposables. Then the one where she spread the word about plant-based diets: even eight meals a week without meat make a huge environmental difference.

But to me, her masterstroke will always be #CutOutCutlery.

Restaurants were automatically chucking plastic cutlery into every restaurant take-out and delivery order. That's food, mind you, that people consume *at home*, where they already own cutlery. That's plastic that people throw *directly* into the trash. (No, you can't recycle plastic cutlery; it's too small for the processing equipment.)

This wasn't single-use plastic; it was *zero*-use plastic. And it drove Sheila crazy.

So she undertook a campaign to convince Uber, Grubhub, Postmates, and DoorDash to *change their apps* . . . so that "Include plastic cutlery" would be a checkbox that's opt-in only by default.

I mean, what? Tell a bunch of cocky Silicon Valley tech bros that they should change their apps? Yeah—good luck with that.

But she succeeded. All four companies eventually agreed. Today, you don't get plastic unless you actually ask for it. That's right: Sheila Morovati figured out how to keep billions of pieces of plastic a year out of the landfill.

I didn't know what to expect from Sheila's book. Would it be a memoir? Or a how-to handbook for change?

The answer is yes.

It's an account of how *one person* produced massive systemic changes. One person, who set out without any particular political power, connections, or money. One person, armed with nothing more than some good ideas, some familiarity with sociology and psychology, and gentle persistence. Somehow, she finds the good in people, and designs ways to lower the barriers—reduce the friction—for helping them do the right thing.

But this is also a guidebook for reproducing her success. Each chapter concludes with a set of takeaways—a template that other people can follow on quests of their own. People like you. Like me. Like cynics.

I still think that the people in power should have taken the climate crisis seriously years ago. I still maintain that they should have made much bigger changes much sooner. I still believe that the entrenched systems favor the status quo of waste, consumption, and burning filthy fuel to generate power.

But I just can't get Sheila's story out of my head. She didn't crawl into a cynical hole, or go around muttering under her breath, or spend her evenings spewing toxic rants on Twitter (X).

Instead, she started small and local. She gradually built an army of friends, fans, and followers. She worked within the system to change it—and wound up producing enormous, global changes in our habits of waste.

OK, let's be honest: Maybe Sheila hasn't snuffed out environmental cynicism altogether.

On the other hand, she's not finished yet.

David Pogue

1

HOW IT ALL BEGAN: CRAYONS

PART 1

I am a great cook if I do say so myself. I am the type of person who doesn't measure ingredients but manages to throw things together to create healthy and delicious meals. I am pretty confident about the meals I prepare—that is, until my daughter was born and started eating solids. It seemed that no matter what I prepared for her (unless it was cut fruit), she just refused to eat. I know this might be an unusual way to open a book about being an "imperfect environmentalist," but this is really where my journey into this work began. As my daughter, Sofia, started to enter toddlerhood, her picky eating became a bigger and bigger issue for us. As a new mom, I was on the cusp of desperation. I took her to our wonderful pediatrician and even to a children's nutritionist, who tried to assure me that she was still thriving despite the few calories she seemed to ingest daily. I was not convinced.

All I knew is that if I went to California Pizza Kitchen, somehow a miracle would happen each and every time we went—she would eat (a lot). She would eat every last bite. To be clear, she ate the same thing each time: buttered pasta, edamame (soybeans), and a side of avocado. I wish I could just express the relief I would feel each time this "feast" would take place. It was just pure joy to watch her take these huge bites of food with zero cajoling necessary. Of course, looking back on all of this, I would like to tell my younger self that I needn't worry at all and Sofia will be just fine. Nonetheless, we would frequent California Pizza Kitchen at least twice a week and order the very same kids' meal each time.

Here's where it gets interesting. California Pizza Kitchen is a well-known family-friendly chain restaurant here in the United States, and part of their kid-friendly experience is to gift a four-pack of brand-new crayons plus a kids' activity menu upon arrival each and every time. This is to entertain the children until the food arrives, which is always relatively quick because it is a casual family spot. After the first few times dining there, before we became regulars, I realized that Sofia was using one or maybe two of the crayons for a few minutes and then pushed them off to the side of the table since the food would arrive so fast. At the end of the meal, I noticed something very odd. The person who cleared the table would grab the crayons and toss them in the same black bin that held the food remnants, used napkins, and so forth. How strange? As an immigrant in this country, the memories of my own childhood started rushing back.

You see, I was born in Tehran, Iran, right before the Islamic Revolution, which took place in 1979. Life in Iran was amazing at that time. Many people compared Tehran to Paris, and it seemed like Iran was set to become a major world hub. Things were going great for my parents. They had a wonderful and fulfilled life in Iran with thriving businesses, friends, family, and me, their new baby girl. There were no hijabs worn or huge discrepancies against women like there are today. In fact, it was quite a modern country until the Islamic extremists took power, which would regress the country significantly. The last thing my parents wanted was for their daughter to grow up living under that suppressive regime, which inevitably limited the freedom of all its people—especially women. This led to my parents fleeing the country with a mindset that they would return in a few weeks once things settled down and the Shah of Iran regained control. What they envisioned as a two-week vacation abroad ended up being a forty-four-year separation from their homeland that is still ongoing.

We moved around a lot until we ended up in New Jersey, where my parents started their life over again. They had nothing but the suitcases they had packed for their "vacation" and a small sum of spending money. Ultimately, while the transition was very difficult for my family, I was very lucky to have an overall wonderful childhood. There was a lot I used to wish for, and a new pack of crayons was definitely one of those wishes. I still recall the joy I felt when I finally got my own new crayons. That new-crayon smell was right there with me twenty-five years later as I watched, alongside my daughter, the precious crayons being tossed away. I would watch wide-eyed as the restaurant staff discarded the colorful gems that had so much joy and hours of creativity left in them.

From then on, each time we went to California Pizza Kitchen I saved the crayons that Sofia received. I did that again and again upon every subsequent visit. However, we were now bona fide regulars, so it took no time at all for me to accumulate over fifty crayons, which was a clear sign that this solution of keeping the crayons wasn't sustainable. To make matters worse, I looked at the tables around me full of kids, each with a four-pack of crayons in front of them, all of which were headed to the trash moments later. How come this wasn't bothersome to the other diners or the restaurant staff? A siren blared in my mind that we are teaching our kids to participate in this "throwaway society" at such a young age.[1] If you haven't heard of this term, a "throwaway society" comes from the notion that our society intentionally produces goods that are meant to become obsolete, such as fast fashion, electronics, takeout utensils, single-use plastic, and much to my dismay, even crayons. As a result, American waste is at an all-time high, reaching 4.9 pounds per person per day in the United States.[2] These "disposable" crayons were contributing to this accepted yet unfortunate mindset. Why was zero value placed on the ubiquitous crayon that carried so many memories for virtually all of us, at least for all of us raised in the United States? Hadn't those parents had some experience with these humble art tools? How could they not *see* the problem here?

Would I have noticed this "crayon-wasting norm" had I not been an immigrant to this country? Maybe not, but throughout my life I learned that being wasteful was not an option, partly because we had to be mindful of our expenses and partly because it just wasn't the culture my parents grew up in. It felt like a little thorn in my side had started to grow, and with each visit to kid-friendly, crayon-giving restaurants I felt this thorn impose itself on me more and more. Why did I have such an issue with these measly crayons being thrown away? Maybe because I knew what joy crayons could bring to a child without the financial means to go out and purchase crayons of their own. Or perhaps it was because the news was regularly reporting on the status of overflowing

Figure 1.1. Dining out with my daughter, Sofia Morovati, in 2010. Photograph by Sheila Morovati.

landfills, which meant we had to rethink what "trash" was. *Upcycling* and *recycling* were the terms of that decade for sure.

As I lay in bed at night, I started scanning for solutions to this unexpected crayon problem I had uncovered. I had a vivid memory pop up in my mind from when I was around nine or ten years old and on one of our first family vacations, close to a decade after our move from Iran to the United States. My parents selected Acapulco, Mexico, to get a break from a particularly harsh winter in New Jersey, and it was incredible. Our hotel had the most wonderful buffet breakfast with virtually every tropical fruit imaginable plus a chef who stood there making eggs to order. The hotel had activities for kids on the hour, and one of my favorites was painting a ceramic parrot, which they put in a kiln for me. My parents were expected to purchase our finished pieces, but that was not an option for my family, and I still remember the remorse of leaving my parrot behind.

One night my dad suggested we leave the hotel and see what the city looked and felt like. He had wanted us to experience the reality of

Mexico outside the very touristy hotel, so we went to a locals' spot for dinner. We arrived at the restaurant, and as soon as we walked in the hostess gave me a few little locally made toys to play with at dinner. One of the items I particularly remember was a small handwoven basket with a lid on it. I put several candies from the hostess desk inside the little basket as a treat for later.

The food was great, and we had a lovely evening. On our way out of the restaurant the kind hostess handed me a yellow balloon with a long string attached to it. I held my balloon tight, along with the rest of the little toys in my arm, struggling to carry them all without dropping anything. It took less than one minute of us walking back to the hotel for a swarm of children to come out of nowhere screaming with delight as they pointed to my balloon. They were overjoyed at the sight of this floating bubble and then they saw my toys. I didn't speak Spanish but knew inherently that they wanted these baubles I had. The next thing I knew, my mother was whispering to me in Farsi, asking me to hand everything over to the children because they would have weeks of joy while I would likely forget about these items by the next day. I did what she asked, and the sadness I could have felt over losing my newly received gifts was overshadowed by the feeling of joy at giving back. I can still hear their squeals of delight.

This small moment in time during my childhood was the driving force that led me to do something about the crayons I encountered each and every time my daughter and I went to a kid-friendly restaurant. As solutions came in and out of my mind, I thought the perfect solution was to send all the crayons I could collect over to Mexico.

This seemed to be a great idea, until I started to make some calls and learned how expensive it is to actually ship crayons, especially internationally. Also, who would receive them? What charity could become the home for at least 150 million crayons that restaurants were discarding per year? I needed a different solution. One that would allow for long-term, sustainable scalability. One night as I was racking my brain for an answer, it suddenly hit me. What if each restaurant collected the crayons left behind and then once a month, *the restaurants* could donate the crayons to a local school? That way, there was no shipping, no extra cost, and no extra carbon footprint, and it would be a local community effort that could be scalable in all fifty states. It was genius. It was a win-

win, and the bottom line was that it would be an easy sell since most everyone can appreciate that it is just wrong to throw away a crayon that could still be a treasure to another child. I started to break it down to see how this solution could be implemented.

The first thing I did was list the many issues we were solving by diverting the crayons to underserved schools.

First, I felt that we were teaching small children as young as two and three years old to be part of the throwaway society we find ourselves in today.[3] This was a key issue for me because we can spend hours in schools teaching kids how to recycle and be stewards of the planet, yet they would still pick up subliminal cues from society that we accept the action of throwing away still-good crayons each time we dine out. This message is clearly not sustainable, and teaching children to upcycle their crayons could be a great lesson to counteract the endless consumption mindset we have found ourselves in.

Second, crayons are made of paraffin wax, which does not decompose because it is a by-product of crude oil and nonbiodegradable.[4] The disposal of these crayons contributes to not only a culture of waste but also the devastating environmental effects of that waste. In other words, crayons are made of plastic, and they will just sit in our landfills for hundreds of years. Meanwhile, many of our landfills around the United States are already overfilled, and as a response to this, cities like Los Angeles are transporting trash to newer landfills.

Third, teachers in the United States are expected to pay for classroom supplies with their personal earnings. Around 2012, it was about $600 per year; today it's closer to $900 per year on average.[5] Over time, teachers have taken on the burden of outfitting their classrooms with supplies due to budget cuts in our education system. How this came to be a solution for already-underpaid teachers is something I have yet to accept, but the truth is, I imagine, that dedicated teachers simply could not do their jobs or, better yet, do their jobs well without the much-needed school supplies, so when the district denied their request for supplies, they dug into their own pockets. Simply put, this is absolutely unfair and unsustainable, especially in a country like America, which has been the richest country in the world for the last sixty years, with an overall net worth of $145.8 trillion and a gross domestic product of $25.46 trillion. Crayon donations could ameliorate these out-of-pocket costs for teachers, solving two problems at once: reducing waste while also supporting both teachers and students.

Fourth, art education was being and continues to be eradicated in public schools due to budget cuts.[6] Over and over research proves that children with access to art education perform better in all subjects and have a higher chance of getting a high school diploma, going to college, and becoming successful adults, yet still we see budget cuts in art education. This divestment in education demands creative solutions that allow for the continuation of programs in the arts. These programs offer powerful learning experiences for students and create space for creativity, exploration, and connection. Underserved communities typically bear the brunt of these cuts. The crayon donations help provide much-needed supplies and reduce costs for programs.

Fifth, the problem was bigger than we thought. Restaurants were throwing away a whopping 150 million crayons per year.[7] We believe it was much more than this since Crayola alone produces three billion crayons a year.[8]

I had to do something to try to repair this broken and wasteful system, so I began by calling all the family-friendly restaurants I knew and asking to speak to their managers. I learned that it was best to call outside of the breakfast, lunch, and dinner rush so they would be more willing to listen. Using the power of storytelling, I would explain my idea that the crayons kids left behind were precious to other kids nearby and we wanted to donate them. Most of the managers thought it was a great idea and would share that they all felt guilty about throwing away still-good crayons but had no idea what to do with them otherwise. At this point I realized that the crayon is a nostalgic item that all of us have some experience with and it's understandable that a child in a vulnerable community would want to have crayons. So I convinced about ten restaurants to start collecting their crayons. The next step was to find a home for all these crayons.

This part was more difficult than I had imagined. The schools had never heard of this idea of receiving gently used restaurant crayons, so they assumed that my call was to sell them crayons. I patiently clarified that I was simply trying to understand if they were in need of crayons at all and that I was facilitating a donation of crayons collected from the local restaurants in their neighborhood.

I spent hours on hold as the administrators asked different departments if they could accept the crayon donation. Sometimes it took calling two or three schools until I found one that would take the crayons. I realized that the schools I needed to call were called Title 1 schools, schools where at least 40 percent of students come from low-income families.[9] I found a great resource online called USA.COM, which lists the name and phone number of every school in the country. It is a great resource, but the task was very time-consuming because I had to click so many times to find out whether the school was Title 1 or not. So I reached out to the website and explained what I was doing and requested that they list the Title 1 standing on the first page along with the name and address of the school. To my surprise, they willingly changed the code on their website to make it easier for me to search for schools. This was a game changer. The time it took me to find the right school was cut in half, and I learned my first and biggest lesson. You'll never know how easily you can receive what you need unless you ask for it.

After the first month I had each of these ten restaurants collecting the crayons left behind, and once per month I did my rounds to pick up the crayons. I was stunned. Each restaurant had collected approximately twenty-five hundred to three thousand crayons. I picked up a large paper grocery bag filled to the brim with perfectly like-new crayons. Next, I drove the crayons to each recipient school, and their jaws dropped too. No one could believe that with a few phone calls we could generate this many crayons—for free. So I started to call more restaurants and more schools. It became very clear that this idea would work.

Suddenly my list grew to twenty restaurants collecting crayons all over Los Angeles, and I realized very quickly that having me deliver the crayons wasn't going to be a long-term solution, nor would it be scalable. So I asked one of the schools that had received crayons from us whether or not they would consider picking up the crayons from their local restaurant one time a month. They said, "YES!" with so much enthusiasm that I felt their sense of gratitude for my having made it possible for them to suddenly have all these free crayons. It was apparent that restaurant crayons were really of value to them, and it was also apparent that this could no longer be a one-woman show.

I decided to jump ahead and call the corporate offices of each restaurant so that I could get every single location within the chain

Figure 1.2. Pile of crayons collected from local restaurants. Photograph by Sheila Morovati.

on board all at once. I called the person in charge of sustainability or marketing to see if they would be willing to participate with all of their locations if I promised to find a match school for each restaurant. But they said no. Not because it wasn't a good idea; in fact, they loved the idea. Rather, the problem was that unless I was with a nonprofit organization, they didn't have the authority to work with an individual like me. As you can imagine, I wasn't ready to give up yet, so I pivoted to my next step, which was to learn how to start a nonprofit. This was quite the undertaking and I realized quickly that starting a 501(c)(3) nonprofit organization isn't for the faint of heart. I felt overwhelmed by just the amount of paperwork (and I am not one to be overwhelmed easily). A friend of a friend who was also a lawyer said she could try and help me on nights and weekends, but soon after she attempted to take this mammoth project on, even she was overwhelmed by the amount of paperwork. There is no question that I was seriously disappointed, but one thing that I learned about myself through this journey is that I don't take no for an answer. Then I realized there was a parent in my daughter's preschool class who was an attorney at one of the largest and most prestigious law firms in Southern California.

His name was Mike, and one morning after we dropped off the kids, I explained the conundrum I was facing. I shared that I was trying to collect and donate crayons on a bigger scale, but the restaurants' corporate teams wouldn't work with me unless I was part of a nonprofit organization. I explained the potential of millions of crayons being saved from landfills and brought to vulnerable children so that they could learn creatively. I went on to say that the restaurants loved my innovative solution, but that I needed 501(c)(3) status in order to work with them. He appreciated my idea and my effort and said that I could apply for pro bono support through his law firm. If I was selected, then they would complete the necessary steps and paperwork to get my nonprofit going. We submitted the application together and before long were approved, and we became a pro bono client of the infamous Irell & Manella law firm. With no issues at all they quickly processed everything we needed for my new nonprofit, Crayon Collection, to be born. Essentially, starting a nonprofit is like starting a corporate entity but with the added layer that you must have a governing body (e.g., a board of directors) watching over your business activity. There are different insurance policies needed, plus different tax documentation to ensure that the money received from donors is tracked and goes toward the program you've committed to in your mission. To this day I believe that running a nonprofit is very much like running a regular business only we must raise money in order to pay for our expenses, such as employee salaries, rent, computers, and so on. Luckily, we also receive many in-kind donations—goods and services that companies are willing to donate to support our cause. If by some chance we raise more funds than we allotted for in the budget, then we are able to expand our reach, which is always a great celebration.

In the short time since beginning my journey into this charitable work I realized the goodness in people and how those around us are our allies, especially if you share your mission and your goal in a compelling way. Luckily, I was and still am deeply passionate about my work, so I don't have any difficulty sharing my story. Normally I manage to inspire others to join in. It is surprising to see how much support is out there and ready to catapult you forward if you just ask for it.

By 2014 I had the 501(c)(3) nonprofit status under my belt, so I felt like I could take on the world. I started calling every family-friendly restaurant chain you can imagine, Denny's, IHOP, California Pizza Kitchen, Cracker Barrel, Buffalo Wild Wings, Islands, and more. Some were easier to convince than others, but the first restaurant chain to join us with every single location was Islands restaurants, located right here in Southern California. In one of our conversations, they estimated that they purchased a total of six million crayons per year. Suddenly I had eighty new restaurant locations and six million crayons to give away to schools in need of supplies. I felt overwhelmed working on this alone, but it was also exhilarating to see the impact I had already made in such a short time. I was so happy to have this challenge and couldn't wait to donate millions of crayons to vulnerable kids. I began by tackling their list of restaurant locations one by one and finding a match school locally that was in need of crayons. I edited my pitch so that the first thing I said was, "I am not selling crayons, but I would like to donate some to you. Do you need any?" This helped us reduce the duration of our conversation so that I could move on to the next school in search of the right match.

It took me about six weeks to pair up each of the eighty Islands restaurant locations, and it left me feeling eager to convince the next restaurant chain to participate. I thought I would easily be able to convince California Pizza Kitchen to join in since the story started with them, but so far I have only been able to convince a handful of their restaurants. But Denny's corporate said I could work with their corporate locations! So I started pairing up about four hundred of their locations. In an effort to get through the list more efficiently, I went back to my alma mater and asked for some help from a sorority whose values were based on giving back. Each Friday morning I would show up to UCLA with bagels, cream cheese, juice, coffee, and fruit and a small piece of paper listing five Denny's restaurants for each student to take on. Their job was to contact a local school within five miles of the restaurant and ask if they'd be willing to pick up the crayons that the restaurant was collecting. After the five restaurants were paired up, the students would get a new piece of paper with the next five restaurants to pair with local schools. We started by calling East Coast restaurants first since they are three hours ahead and would just be finishing their breakfast rush, then we would focus on the Central Time Zone, and end with the West Coast. This was before Google Docs started being used for collaborative editing, so one student's job would be to enter all the school and restaurant matches in an Excel file so that we could keep track of every pairing.

Once I realized I had this army of students helping me, we started on all the Buffalo Wild Wings nationally. At this point, my nonprofit consisted of me and the help of many volunteers, and no one was getting paid since we hadn't even started fundraising.

UCLA let out for the summer and we had no one to continue the work we had started. Suddenly I was faced with my first major obstacle. I received one email after another from the restaurants that were collecting crayons all summer, but no one was picking them up. Of course, the Title 1 schools were all on summer break. What was I going to do with hundreds of restaurants with crayons ready to be donated? I realized that the National Head Start Association might be a great organization to partner with. They are a division of the U.S. Department of Health and Human Services that handles all the publicly funded preschools, which serve the most vulnerable population in the country. I began researching their mammoth organization to find the right person to speak with. I found someone in Programs and Partnerships at the National Head Start Association who happened to also be a former teacher. I explained that I had thousands of crayons ready to be donated to her students. Given that she was a former teacher, she immediately appreciated the concept of Crayon Collection. She was in awe and understood why I was calling. The work was simply too much for just one person and my student volunteers. So this incredible hero, a former teacher, saved the day. She asked me for the list of restaurants that were ready to donate crayons, and she was able to map the closest school to each location. Next we sent out emails to each of these preschools to see if they wanted crayons. Problem solved.

By this point I had uncovered the key to scaling the Crayon Collection model nationwide. I was on my way to redirecting millions of crayons from landfills and into the hands of underserved children at Head Start centers and Title 1 public elementary schools.

The next challenge was uncovered when I learned that most public schools in America no longer offer art education due to budget cuts. I knew Crayon Collection could help resolve this unfortunate reality as well. These humble crayons had much more life in them than just as a coloring tool.

PART 2

Often I tell people that as you move forward through the motions of a project that you are passionate about, other doors and ideas open that you would never have seen coming otherwise. As I started to get a handle on the process of collecting and redistributing the crayons, I visited many schools and shared how excited I was that the school would now have more crayons for their art programs. I was quickly corrected by administrators explaining that due to budget cuts in education, many public schools no longer had the funding for art education, so it had been removed altogether. This was not an isolated phenomenon. Following the 2008 recession, over 80 percent of schools experienced reduced budgets, frequently resulting in drastically reduced arts curricula.[10] I had a solution for this.

My first step was to try and build an even bigger system so more teachers and schools could receive free crayons from local restaurants. This is when it dawned on me that *any* school *anywhere* could be receiving crayons if *the schools* were the ones making the request from their local restaurants. This meant that any teacher or school could follow our steps to ask a local restaurant to collect the crayons kids leave behind, and the schools would pick them up just in time for the new school year. This helped propel us forward by having thousands of teachers reach out to their local family-friendly restaurants. Restaurant managers were more than happy to support teachers directly by donating the left-behind crayons. The directions we provided were simple. Teachers were instructed to visit their local family-friendly restaurants and ask the managers if they would be willing to donate the still-good crayons left behind by kid diners. The idea caught on and teachers who received the free crayons started posting about it on Facebook and Twitter and spreading the word to other teachers.

This reverse process was brilliant and scalable because it didn't solely depend on me to create the connections between schools and restaurants. It made perfect sense for schools to use the Crayon Collection model locally, which was even more powerful than my nonprofit organization based all the way out in Los Angeles making the request on behalf of the school. Plus, this simple addition to the Crayon Collection program opened the possibility of so many more schools helping us divert crayons from landfills, and so many more classrooms began benefiting from free and still-good crayons. After all, we were addressing the wasting of at least 150 million crayons per year, which we still believe is an understatement.

Next, I needed to find a way to expand our reach so that more people would know that this idea exists. Without the help of a publicist or anything that would cost money, I found the company that runs the National Calendar. I am certain many of you have heard of #NationalIceCreamDay or #NationalDonutDay or #NationalSiblingDay. Well, there is a company that manages all these nationally recognized days. I called them up and made my case for Crayon Collection to be on the National Calendar and they loved it.

Instead of requesting #NationalCrayonCollection day, I made a case that we needed August to be #NationalCrayonCollectionMonth so that teachers could benefit from the whole month to generate crayons for back-to-school. With this lead time, restaurants could produce around three thousand crayons, which is enough for an entire elementary school.

We deemed August as National Crayon Collection Month, and still to this day, every year publications cover the story of Crayon Collection nationally with a call to action for people to participate in receiving free back-to-school classroom supplies from their local restaurants.

Once I knew there was an easy way for any school to receive crayons, I decided to try to tackle the unfairness of the lack of art education. For me personally, this was very bothersome. Some of my happiest moments in grade school involve me zoning out completely while being immersed in a creative art project. I looked so forward to those art classes where I could just let my mind go and start to feel a deep sense of peace. Nowadays people call it a sense of flow.

In addition, many studies are published each year proving that children with access to art education perform better in all subjects. For example, one study found that 14 percent of students who took art history performed better in English classes, 20 percent in math, 16 percent in science, and 16 percent in social studies.[11] The more I researched, the more it became clear that schools around the United States were being denied art education resources despite proof that children with access to art education were more successful in the long run and had higher high school and college graduation rates.[12] If the arts could give the kids we served a chance to break the cycle of poverty they were in, then I felt we should focus on creating a change. All in all, I knew it was simply an awful thing to rob our children of creative learning.

My solution was to challenge professional artists to help us. These weren't just any artists. These were artists with master's degrees in fine arts who had gallery representation, and some even had museum exhibitions. They were the true definition of an artist, and I knew that they would like to give back and help bring art education back to vulnerable children. Luckily I was able to ask a few local galleries, such as Honor Fraser Gallery in Los Angeles, to pass the request on to their artists.

The request was for them to share what they would create using just crayons as the main tool—nothing else. I told them to imagine a child that had never had any exposure to art education, and thanks to these salvaged crayons and the art project they were creating for us, we could unlock a whole part of them that would have never come out. The results were absolutely astonishing. Seeing how an artist's mind works is still incredible to me. This became the Crayon Collection Arts Education program.

The first project I received was by an artist named Annie Lapin and it was called "Newspaper Shapes." If you ask me, she is the definition of a true genius. She created a project using old pages from a newspaper because she took our directions literally when we said the only tool the child would have access to would be a crayon, so she went so far as to provide newspaper as the replacement for regular paper. (We have since changed our instructions when we challenge artists so that it allows for paper.)

Annie's brilliant project started by having the child circle the first letter of their first name all over the page. They would then draw lines to connect these circles. From there, they would begin to see a large

By ALINA TUGEND

It won't be the old car or the thousands of tobacco leaves or even the leaping crocodile model that will greet visitors when they walk into "¡Cuba!," the first major exhibition the American Museum of Natural History has ever organized about that island nation.

Guests will be [illegible] by life-size photographs of Cubans from Cuba and New York [illegible]

[illegible]

"¡Cuba!"

"We wanted the exhibition [illegible] honest," he said, [illegible] Cuba, [illegible] to come through.

"¡Cuba!," [illegible] through [illegible] est [illegible] natural [illegible] presented in the [illegible]. It will also be the [illegible] first fully bilingual show.

All the senses will be brought into play. The scent of tobacco leaves will waft through part of the exhibition. Videos of Cuban music — and perhaps even live performances — will play at the end of what looks like a typical Cuban street, with cafes (no food or drink, but guests can sit down and play a game of dominoes to the aroma of coffee).

The timing is both serendipitous and not. It comes about 16 months after President Obama announced the re-establishment of diplomatic ties, which were severed by the Cuban missile crisis more than 50 years ago. Legal commercial flights to Cuba began just this summer.

But the museum's ties to Cuba go deep. "We have a relationship with Cuban scientists and paleontologists going back 100 years," Dr. Raxworthy said.

While the museum had long had an interest in doing something on Cuba, "the normalization awakened people's interest," he said. The political shift also made [illegible]

[illegible]

Nothing is being shipped from Cuba because of the tight controls on the movement of biological materials. In any case, Dr. Porzecanski said, it is common for the museum to make models; the exhibition needs to be hardy enough to travel, as it might be shown in different museums over the next 10 years.

So in the bowels of the museum, an entire kaleidoscope of Cuban life has been created.

It took 25 volunteers and several staff members to make the 4,000 tobacco leaves out of construction paper. They are hand-painted with acrylic in varying shades of brown and green, complete with tiny holes mimicking insect damage. Stems made of hardened hot glue are attached. Then they are crinkled, bundled together and hung over dry[illegible].

"It's taken [illegible] April to the end of August" to [illegible] Andrea Raphael, a [illegible]

[illegible]

[illegible] original posters created over [illegible] decade, which show "the breadth [illegible] of Cuban art," said [illegible] director of exhibition graphic [illegible]. Michele Miyares Hollands, a Cuban artist who created some of the [illegible], was brought in to help develop that part of the show.

An interactive exhibit allows museumgoers to explore the country's art, fashion and dance worlds. A 1955 Chevy will be on display as a tool to explain the embargo against Cuba that prevented most new cars from entering the country, as well as the ingenuity Cubans used in keeping their ancient automobiles running.

But of course, the natural world is an important part of the exhibition, and that means an enormous amount of work has been done to recreate accurately everything from the prehistoric giant owl to the bee hummingbird, the tiniest bird in the world. Wetlands, rain forests and coral reefs all come to life.

The plywood, foam and resin model of the Garden of the Queens coral reef alone took about six months to make, complete with a 12-foot shark, which once lived in the museum's Ocean Life hall and had been banished

Figure 1.3. Students create the "Newspaper Shapes" project curated by artist Annie Lapin for the Crayon Collection Art Education Program. Photo provided by Crayon Collection.

shape coming together. This shape would direct the next step, which was to start making the shape into the animal the shape reminded them of. Next they could color the image in and add any more elements they felt would enhance the final product. In the end the project was different each time since the newspaper articles would have different text and thus different places to make the circles. The child was left with a beautiful project that had inspired creativity and was absolutely free.

I could hardly wait to share this new art education program with all the schools we served. My expectation was that they would be as happy and excited about these projects as I was, but alas they were hesitant. I couldn't believe it. What was the problem? We had provided free crayons, and now we even had free art education resources. The teachers were unwilling to incorporate the resources into their classes because they were overworked and underpaid, and understandably they truly did not have time "to teach art too." I had clearly witnessed that so much was already being asked of teachers and that teaching art on top of everything else was not helpful or possible.

With more compassion for educators than ever before, I found a retired teacher to build the art projects within the guidelines of the standards teachers must teach anyway. For example, Annie Lapin's project was applied to the Common Core standards for Language, Reading, Speaking, and Listening, in addition to the following Visual Arts Standards: Artistic Perception, Creative Expression, Aesthetic Valuing, Self Awareness, and Social Awareness. We created lesson plans with all new codes and verbiage covered in three-page documents for each art project. This gave me hope that teachers would incorporate some creativity into their classes. It worked! Since we transformed the art projects to be compliant with Common Core Standards, which teachers have to teach anyway, they were much more well received. Art education became the by-product of all of our sustainability efforts—or, as I like to call it, the cherry on top.

I thought it would be meaningful to invite the artists to teach their project ideas inside the classrooms that were around the Los Angeles area so they could teach the class themselves and interact with the students. The children we served had no idea that they could grow up and pursue art as a career. They had never heard of such a thing and thought it was a fabulous idea. Going into these classrooms gave me, and the artist volunteers, perspective on what the crayons meant to the kids and how powerful it was for the children to hold these colorful gems. We ended up receiving a best-in-class award from Los Angeles Unified School District for our art education program and crayon recycling model.

The classroom visits inspired me to try to raise even more awareness for teachers and to tell the world that restaurants were discarding over 150 million crayons per year that could go into classrooms everywhere instead of landfills. I decided to set a Guinness World Record, which if I am being honest was probably the hardest thing I have ever done. We collected 1,009,000 crayons and displayed them in the shape of a maze created by artist Yassi Mazandi. Once the counting was done, we distributed every last crayon to nine hundred teachers, who had been waiting over three hours in line. They all worked within the Los Angeles Unified School District. Our goal that day was to prove that crayons don't belong in the trash and that it is important to really understand what we are doing instead of blindly moving through our days unknowingly impacting the world around us negatively.

In-classroom visits continued to be a highlight of our work. I regularly asked the students if they had crayons of their own at home and most said no. Many said they never owned even one crayon. So once again I decided to create a final Crayon Collection program called the Color Kindness program. I was still keenly aware and distraught that many children dining out mindlessly leave their crayons behind and are missing a very valuable lesson in conservation. They are partaking in the throwaway society at such a young age without even realizing it. So I started asking well-served schools to start Crayon Collections of their own and had the kids in these schools do a service learning activity that involved collecting crayons and then packing individual crayons (about ten per pouch) along with a small note or drawing of kindness (more on that later). These packs would then be donated to children around the world, including in Uganda, Mexico, Guam, Liberia, Haiti, China, Guatemala, Iran, and more, thanks to the many people and organizations who were able to transport these on their trips. Many times, these little packs of crayons were the first ones a child could call their own. This was imperative to me because a child can escape many struggles with their crayons; they can imagine and create, and they can develop so many fine motor skills.

Project conceived by artist Pearl C. Hsiung for a collaborative workshop with Crayon Collection at the Institute of Contemporary Art, Los Angeles (ICA LA) on March 10, 2018

"Beyond Our Bodies and Into the World"

Figure 1.4. Crayon Collection Arts Education project titled "Beyond Our Bodies and Into the World," conceived by artist Pearl C. Hsiung. Image courtesy of Crayon Collection.

Figure 1.5. Crayon maze featuring 1,009,500 crayons created by artist Yassi Mazandi and Crayon Collection for setting a Guinness World Record. Photograph by Sheila Morovati.

I realized that more connections could be made if the kids donating the crayons could spend a few moments to be present and include little notes of kindness inside the crayon pouches—something as simple as a few positive words or a drawing. Since the inception of this program we have seen thousands of notes of kindness written by one child to another, and it is so inspiring to see the care and love that children have in their hearts. Our Color Kindness program has since been adopted by countless Girl Scout Troops and even corporate volunteer days at enormous institutions like Microsoft, Hulu, Netflix, Nickelodeon, and others. My favorite was seeing very serious-looking Microsoft engineers handling crayons by packing them in small pouches, plus writing notes of kindness as part of their company service day.

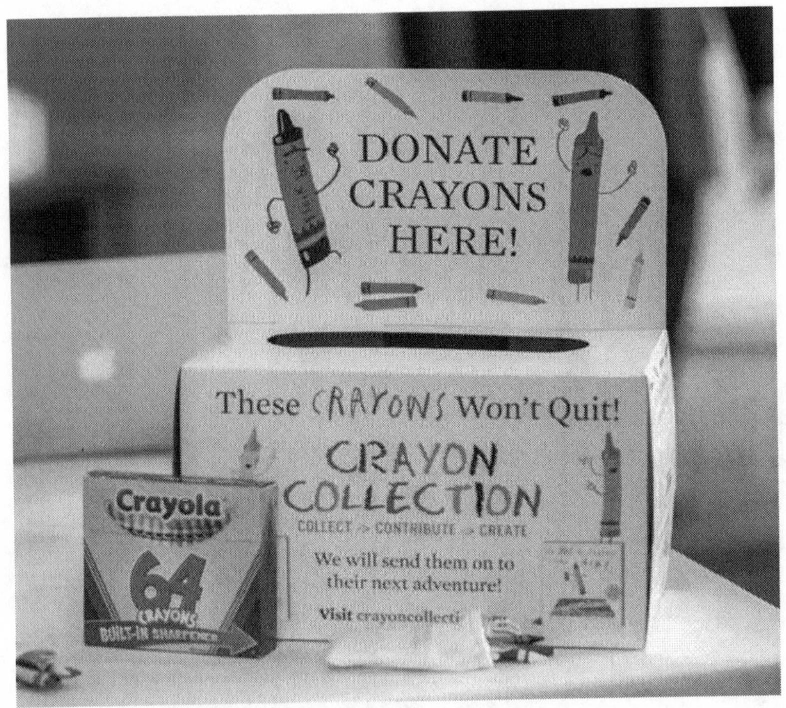

Figure 1.6. Crayon Collection donation container. Photograph by Sheila Morovati.

Still, the most special notes are those of kindness written by kids to other kids. It seems to bring out the best in them and displays their innate ability to show love and compassion.

The Crayon Collection Color Kindness program caught the attention of many schools, companies, and even museums, so we had kids and adults all over the world collecting and donating crayons and drawing kindness notes. Both the crayon-donating kids and the recipient kids connected through these notes of kindness, and both sides benefited greatly. In fact, studies show that children who volunteer their time to charity are actually happier.[13]

Every time we visited a school our focus was to provide each student with a pack of their own crayons to keep. The beginning of class was devoted to opening the crayons and reading their notes of kindness. Many times, I would notice the kids hold these notes to their hearts with big smiles on their faces. Then came the big surprise. We would announce that the crayons sitting on each child's desk were theirs to keep. Never had I heard such loud screams and squeals of delight. They

Figure 1.7. Student art created for the Color Kindness program. Photograph by Sheila Morovati.

simply could not believe that this little pouch of crayons was theirs to keep. It was a further testament to why we should not be throwing still-good crayons away.

The year 2024 marks the ten-year anniversary of Crayon Collection becoming a registered 501(c)(3) nonprofit organization. Millions of crayons that were headed for landfills have been diverted into the hands of thousands of children worldwide. These days we see the systems change that has developed within restaurants. We notice how willingly they will save crayons from entering landfills, allowing us to teach environmental stewardship, kindness, and art education. The innovation of the Crayon Collection program opened many doors for us over the years, enabling us to create partnerships with the Getty Museum and the Institute of Contemporary Art in Los Angeles, hold events in conjunction with Los Angeles County Museum of Art, and work with countless universities and corporations big and small, including Microsoft, Penguin Random House, Nickelodeon, Hulu, and many more. We have been advocating for art education for all since 2014 and were so glad to hear that Prop 28 passed in the State of California in 2023, reinserting $900 million to fund art education resources in all schools across the state. The thought of crayons being thrown away hits a nerve throughout society because we all share some connection to the beloved crayon.

Our most recent campaign is called "Save the Crayons" where our supporters can text CRAYONS to 52886 and it automatically sends an email to major family-friendly restaurant chains, including Denny's, IHOP, Applebee's, California Pizza Kitchen, and Cracker Barrel, convincing them to donate their crayons to local schools, instead of discarding them, using the Crayon Collection Recycling program. In addition, we thought we would start trying to create laws that would stop the insanity of throwing away still-good crayons. For that effort individuals can text CRAYONLAW to 52886, and upon entering their zip code they are automatically put into touch with their local senator or assembly member requesting a ban be placed on discarding still-good crayons.

Now that you have read this chapter I can guarantee one thing: You won't ever look at a crayon the same way again. Many say that they cannot "unthink" it and feel compelled to speak up and help. We hope you do the same.

What this little crayon did for me after starting Crayon Collection will fill the pages of this book. I was inspired to look more deeply at our daily habits to see what else we could be doing differently. I knew there were more win-win opportunities out there to seize. That would become my mission moving forward.

HOW YOU CAN BE AN IMPERFECT ENVIRONMENTALIST

1. Anytime you see a restaurant giving out crayons, let them know about Crayon Collection so they can start collecting crayons for a school in their neighborhood. We can help them if they reach out to us via www.crayoncollection.org.
2. Get your kids involved. Set up a Crayon Collection box in your child's school so that kids can easily donate the crayons they receive when dining out or those they no longer need at home.
3. Speak out if you see something that seems wasteful. You never know what will happen if you just share your thoughts with a few people. Chances are that you will find supporters out there to help you get your ideas heard.
4. Text CRAYONS or CRAYONLAW to 52886 to make your voice heard by restaurants and lawmakers. (Direct link is also available at www.crayoncollection.org.)
5. Never shame or blame anyone into listening to your ideas. Be open-minded. People are more willing to change their ways when they are a part of the conversation as opposed to being told they are bad and must change.
6. Use social media to help you. It's a wonderful tool to get the word out and share ideas for more eco-conscious living.
7. Let your local schools know that they can receive free crayons from local family-friendly restaurants. All they need to do is ask!
8. Become a Crayon Collection Ambassador so you can continue to spread the word about our work in your community.

9. Start a Crayon Collection drive in your local church or community recreation center during National Crayon Collection Month (August).
10. Spread the word that crayons are not trash! With our solution many children benefit deeply from having access to these colorful gems.

2

STRAWS AND MALIBU

I owe a lot of the work I do today to an anthropology class I took at UCLA. It was one of those classes that you take to fulfill the requirements of your major; mine was sociology. I went into the class with virtually no knowledge of what it meant to be an anthropologist. The professor was amazing. He was from Nigeria and had the most interesting accent, which made me hang on to his every word. The first assignment we had was to go to the student cafeteria and observe the way people eat—not just how they chew and physically eat, but also their rituals and processes. It was the first time I became an observer. I decided to perform my anthropological foray in the morning, and it seemed that many students started their mornings out with these huge muffins for breakfast. After about an hour of observations I noticed one similarity among the men: they all took a bite straight off the top of the muffin. The female students never did that; instead, they took a piece with their hand and placed the fluffy, bite-sized portion in their mouth.

I wrote a paper on these observations, got an A, and completed that class with the realization that by observing virtually anything or anyone you begin to notice a pattern that would have been so discreet otherwise. It is very clear to me that I would have never noticed these nuances of human behavior had it not been for this assignment. To this day I often see the same muffin-eating behaviors between men and women.

Looking through the lens of an anthropologist helped me "see" all the crayons being tossed in restaurants. It also gave me the insight to notice another habit of wastefulness that was happening even more often than the tossing of crayons: plastic straws. At this point my daughter (Sofia) was about six years old and I was tackling the issue of crayon waste as my daily job. We were still dining out at family-friendly restaurants often thanks to Sofia's ongoing disdain for my cooking. Crayon Collection had been running smoothly for about three years at this point. It was the spring of 2017, and I was passionately telling the world about my work and how every kid-friendly restaurant should be adopting the Crayon Collection model as a way to give back and limit the unnecessary act of discarding still-good crayons.

These preliminary years made me so much more aware of the extreme level of wastefulness everywhere, so I began watching closely to see more mindless behaviors or habits that unknowingly hurt our planet. It didn't take very long to notice a huge problem. Every time I dined out, I observed that the beverages on every table—glasses of water, soda, coffee, iced tea, Arnold Palmers, you name it—all had one thing in common: a plastic straw inside the glass. Sometimes a table of two diners had six straws by the end of their meal because the first glass of water that came to the table upon arrival had a plastic straw, then the diner may have ordered a soda, which also came with a plastic straw, and then of course there were the refills that came with, yet again, another plastic straw. Let's not forget the coffee at the end of the meal, which came with, you guessed it, a mini plastic stirrer.

I wondered if anyone had asked for these straws. How had plastic straws inserted themselves (no pun intended) into our society in such a way that we could no longer expect to have a drink without a straw? It seemed that we no longer could fathom the idea of lifting the glass to our lips to sip the drink. I was fascinated by the evolution of this real dependency we now had on plastic straws.

Earth Day 2017 was coming up, so I prepared to write a blog for the small list of subscribers I had amassed for the Crayon Collection newsletter. It was an inspiring piece that helped people see for themselves that we must do our part to protect our precious planet while always putting pressure at the top too. I wrote about all the "habits of waste" that were happening around us, such as the crayons we were mindlessly throwing away and the single-use plastic straws. I wanted to find a stat, so I did a little research and learned that five hundred million plastic straws were being discarded per day in the United States.[1] Of course, I assumed it was incorrect because *five hundred million* (!) a day seemed to be too high. Alas, that stat was everywhere. I included the information about the plastic straws in the newsletter, and the number of people who wrote back to me was incredible. I had struck another nerve. Many people had no idea about the negative impact these pesky plastic straws have on our environment. A very visible effect of plastic straws is marine plastic pollution, but it is not the only negative environmental impact of plastic straws. From greenhouse gas emissions during plastic production to pollutants produced through end-of-life disposal, the entire life cycle of a plastic straw is fraught with destructive environmental harm.[2] Interestingly, most people wrongly assume that plastic straws are recyclable, but they are too light to be recycled and have no more value in them after one use. We are polluting our planet one drink at a time, and the plastics industry continues to lead us to believe that we can use as much plastic as we want and recycle it endlessly, which of course is untrue. The overwhelming response to this 2017 Earth Day newsletter is one of the first reasons I wrote this book. People care. And they are interested in creating change.

Below is the newsletter I wrote; its responses ignited me to spearhead the world's first plastic straw and cutlery ban.

Earth Day: A New "Eco-Normal"
From the founder of the Crayon Collection

I recently attended a dinner party and was seated at a table with an astronaut who had been to outer space twice in his lifetime. He shared what his experience was as he left the Earth's atmosphere and looked at our planet from above for the first time. Interestingly, he wasn't surprised seeing Earth from the window because he has seen images of the Earth from above many times before. What did take his breath away was the greenish halo that surrounds our planet. The ratio of this halo to the Earth was described as the equivalent of the skin of an apple. Minuscule and delicate . . . very delicate. This halo is us; it is our atmosphere. This is where we live, breathe, eat, sleep, work, love, learn, fight wars, make peace and sadly where we are destroying our planet. That view is what brought this astronaut to realize that if it weren't for this halo, the Earth would be one giant dead rock. Preserving that halo is our job.

Imagine someone were to ask you to eat pebbles and stones in hopes that you could digest it. That your body could withstand the removal of a few organs here and there, and that you should just keep going as though you are whole, without missing a beat. You see, to me, the land we live in is just an extension of ourselves. How can we expect our planet to maintain its loyalty to us day in and day out when we are mindlessly destroying it? We are asking our precious Earth to digest billions of pieces of plastic and other toxic materials, which of course is impossible. We are cutting away at its organs by cutting down its rain forests and polluting its oceans. Despite popular belief, this is not someone else's problem. It's everyone's problem and we can all do our part by taking some basic steps in the right direction.

There are some things that are a part of our daily norm that we may not even notice anymore but that contribute to the disastrous path our planet is facing. Have you noticed that when you order a drink at a restaurant it automatically comes with a straw? Well, that straw is one of 500,000,000 straws that are dispensed each day, never to decompose, poisoning sea life. Or have you gone to a kid-friendly restaurant and received a four-pack of brand-new crayons? Often, they are left behind destined for the already full landfills, never to decompose. (Luckily, we found a solution for the 150,000,000 wasted restaurant crayons, but we still have a long way to go—visit crayoncollection.org to learn more.)

I just came back from the Coachella Music Festival. I stood in one spot for a few moments and watched a portion of the 100,000 different faces walk by me eager to get to the next stage. These faces I saw were all different shapes, colors, sizes with different likes, dislikes, dreams, wants and wishes. Every single one of them had this in common, we are all living on this planet together. We are like neighbors in a very large neighborhood—we are one.

Scientists are saying that it is this next generation and possibly the one after that will "make or break" the state of our environment. They are the ones who can determine the trajectory of our precious planet. It is our job to teach our children about environmentalism from an early age and plant the seed to create a new "eco-normal" by taking notice of all the regular things that are contributing to our planet's demise, such as plastics and carbon gases. Consider making a small change in your daily life—switch to paper straws or better yet tell your waiter you don't need one in your drink. Teach your children about conservation by sharing the Crayon Collection with them—they *love* to get involved. Start a compost with your kids—it's another great way to experience cycles of nature. Lastly, please teach your children about the state of the environment and how much we are counting on them to be eco-minded citizens of the Earth.

Thank you,
Sheila

Another thorn in my side was starting to emerge. In fact, when going to these restaurants I felt like I was in the middle of a barrage of trash that I was constantly trying to dodge. My common first sentence after hello was "No straw please" or "No crayons please," and as you could imagine, they would still arrive at my table regardless of my pleas. Unlike the commonly discarded crayons, I realized there was no reuse option for plastic straws. But I had to find a way to resolve the five hundred million plastic straws Americans were discarding every day. One thing I knew was that by building awareness we could create a grassroots movement of action.

Well, these plastic straws were not on many people's minds at the time. It was 2017 and the idea of plastic straws being public enemy number one was not part of our culture yet. Several amazing organizations were starting to raise awareness, though. It's probably the reason that I even "saw" the excess straws given out in each restaurant I visited. I'm thinking particularly of the "Stop Sucking" campaign by the nonprofit organization called Lonely Whale. It was filled with celebrities and had a very sexual tone to it.[3] I believe this campaign's starring actor, Adrian Grenier, helped propel the movement tremendously because it made everyday people lean in and find out what in the world were they referencing in their cutting-edge public service announcement.

A buzzing momentum was being built. It was great timing for my next steps, which brought back my background in sociology (it's still a major surprise for people to learn that I didn't study environmental science). I had learned about social movements and what it takes to create a shift in culture and values among the masses. There was a common formula that our predecessors had used; most of the time it began with one visionary leading the charge. I was not planning to be that person, but since I was coming with a totally different standpoint from most of the environmental organizations, I had a fresh approach. It was an approach that tied in beautifully with all the research and awareness the other environmental organizations had generated. So I began to create my own grassroots movement against plastic straws.

My first step was to start sharing this idea with like-minded people in my network, who also began sharing and making introductions. I started talking to any and all of my friends or acquaintances who believed in protecting the environment. Within moments of connecting on a call or in person it was clear that we all shared a similar passion to protect our oceans and our planet. One key introduction was made by my cousin Olivia, who was born and raised in Malibu. She put me in touch with Casey Zweig, who was working in sustainability under the environmental director of the City of Malibu. She took my call, and I asked why a beach community like Malibu still allowed plastic straws to be used in restaurants and cafés. She patiently explained many details about how new ordinances are passed in their city, which was essential information for me. (Many years later, I heard Dr. Jane Goodall say that the first thing to do is to create a group because it will help you move forward. Luckily my first instinct was spot on.)

Figure 2.1. With Jane Goodall. Photograph by Sheila Morovati.

Casey was a wealth of knowledge. She explained that there wasn't an outright ordinance against plastic straws, although restaurants were asked to adopt several best practices that the City of Malibu recommended in order to be more eco-friendly. A seed had been planted in my mind to create a full-fledged ban against single-use plastic straws in Malibu. I knew that this would be the shortest way to a worldwide ripple effect. You see, my father's side of the family lives primarily in Italy, and they absolutely love California, especially Malibu. I would even dare to say that they have a mini obsession with Malibu because in their view it is the most iconic beach city there ever was. I knew that if Malibu made a move against plastic straws, then the world would hear about it, especially my Italian cousins. Coupled with the strong public reaction following a viral video of the removal of a plastic straw from a sea turtle's nostril, I knew that other cities would join in as well.[4] But I had to figure out how to make this happen, as I had never come even close to any city's ordinance-building process let alone attend a city council meeting.

One afternoon I received a call from a new friend who was part of my friend group of anti-plastic-straw confidants. She was a producer of the documentary film called none other than *Straws*, which is based on the entire history and background of plastic straws and how they came to be so prevalent in our society, as well as how damaging they are to our planet after being used for a few minutes in someone's drink.[5] She asked me if I wanted to screen the film with her in a cool spot in Los Angeles like Soho House where people were clamoring to get in. Surely that would be the sexiest spot to screen a film like this and even have a panel discussion. But something inside me rejected this idea. I felt that there was an even bigger opportunity at hand.

I wanted to know if we could screen *Straws* in partnership with the City of Malibu at city hall. By now I had learned that I had three minutes to speak at the public comment segment of any city council meeting. I was well aware that if I used my three minutes of talking time and came out of nowhere requesting a ban on plastic straws, I would likely be met with a "we will add it to our list" type of reply. Instead, I went to the city council meeting and requested that we show the documentary film *Straws* at city hall as a free event for the public. This way we could educate the City of Malibu and its residents with how bad the problem really is, as well as answer the community's questions with a panel after the screening. I felt strongly that this was the right strategy because I knew this was a revolutionary ban that had never been done before and we needed the buy-in of the community in order to support the council members in creating what would ultimately become the most comprehensive ordinance on single-use plastic to date.

The council members agreed to allow the screening to take place in partnership with Crayon Collection, and it was a huge success. Even though it was a free event, it was totally sold out, and people were genuinely excited! We had an incredible panel after the screening that included the film's director and a producer, as well as a scientist and an activist who were in the film. I was the moderator of this panel discussion, which was so informative and rich. The panelists stayed until the very last question from the audience was answered during the Q&A portion. We closed the night with a special performance of the song "Hope" by one of my favorite artists, Joe Sumner, also a Malibu native. By the end of the night there was no doubt in my mind that Malibu was going to be the first city in the world to put an end to the incredibly damaging single-use plastic straw.

The following city council meeting was held about a month later, so I went back to city council this time with my kids and used my three minutes of public comment to make my request that the City of Malibu pioneer the first plastic straw ban in the world. The night was *very* long. I

Figure 2.2. Invitation to the screening of *Straws* in Malibu. Image courtesy of Habits of Waste.

remember there was another issue being discussed that evening, so there were about twenty others waiting to speak during the public comment portion, and my son was falling asleep by 9:00 p.m. The mayor at the time was Skylar Peak, and I will not forget the kindness he showed to me and my kids. He saw my children starting to lose steam, so at one of the breaks he handed us a paper cup full of chocolate M&M's to keep us going. This is a memory I hold on to vividly because it was yet another layer of support that I needed that night to just stay a little while longer and wait for our turn to speak. You see, when children are activists, the decision-making adults are really put in a position where they must listen. Part of it is because we are all aware of how difficult it is for a young child to stand up in front of a room full of adults and have center stage. But more important, when it is time to discuss the future of our planet, all of us know that the children are going to be the ones to deal with the mess we leave behind. There is an unspoken respect given to kids who are trying to create change. I am not suggesting that kids must aim to be legendary like Greta Thunberg, but what I am saying is that if the

Figure 2.3. Speaking with my children, Sofia and Leo, at Malibu City Hall to urge the city council to ban plastic straws. Photograph by Sheila Morovati.

kids in your community get involved, everyone is all ears, and change is much more easily achieved. (Kids also get three minutes to speak during public comment.)

My daughter was nine years old at the time. She used her three minutes of public comment at Malibu City Hall to recount the many beach cleanups she had done through her school and share her findings that plastic straws were among the top trash particles she had cleaned up. She then used her final few seconds to urge the mayor and council members to take action for the sake of our beaches, our oceans, and her generation as a whole. She was so powerful, and it is one of my most proud moments as her mother. My son was only seven years old and to this day is still shy when it comes to public speaking, so he asked me to speak on his behalf while he stood next to me holding my hand. After we used our collective nine minutes of public comment and every factoid, stat, and pull-at-the-heartstrings type of effort I could produce. We talked about the future of our planet and how by 2050 there would be more plastic in the ocean than fish; we brought up the marine life and how damaging plastic straws are to their well-being; we talked about the sunny days that brought the crowds to the beach, which ended up looking like a plastic straw cemetery by the end of the day; and we finished with the fact that the planet we are leaving behind is going to be the home of many future generations. We knew the City of Malibu understood all these concerns, so we empowered them to be pioneers for change. Even Mayor Peak recounted his experience of surfing in the ocean and seeing the problem firsthand. He expressed his frustration about feeling plastic straws in his hands when he paddled out to surf recently. That night, the council members voted unanimously to have the new ordinance to ban plastic straws drafted. Whether they surfed or simply walked along the beach, they all had one thing in common: They were all deeply connected to the ocean and more than willing to help create a historic ordinance to stop the endless stream of single-use plastic that was littering their beautiful beaches. They wanted to be a part of this movement because they knew firsthand how awful the problem of single-use plastic was.

I was invited to meet with the environmental team of the City of Malibu to help draft the ordinance, which is now considered the gold standard because it really is the most comprehensive ban to date. The council members had decided to include not only single-use plastic straws but also single-use plastic cutlery and stirrers in the ordinance. We also included a ban on compostables, also known as biodegradable plastic alternatives, because a little-known fact is that these items behave just like plastic in our environment unless they are properly composted to a scorching 180°F in an industrial composter, which are few and far between.[6] To give you some perspective, the closest one to Los Angeles is 150 miles away.[7]

The ban on plastic straws, cutlery, and stirrers had its second and final reading in November 2017. The day after this reading I was a chaperone for my daughter's school field trip at the Getty Center and my phone started ringing nonstop. Since I had no experience with press coverage and new city ordinance readings, I had no idea what was going on! I told one of the co-parent chaperones, who was a writer for the *Wall Street Journal*, that I had received a voice message from the AP (Associated Press), which made his jaw drop. He said get ready, you are about to make worldwide news.

He was right. Within twenty-four hours there were articles in countries worldwide and in every language you can imagine about the historic Malibu plastic straw and cutlery ban. The world was listening. Even the BBC in the United Kingdom sent a camera crew to interview me and the council members in Malibu. It was a press blitz that I was totally surprised by but totally welcomed. This was exactly what I thought would happen if Malibu took a stand; I just didn't realize that I would be a part of it too.

The best part is that the ripple effect started almost immediately. After Malibu had made it official, Buckingham Palace announced no plastic straws. Europe, China, Australia, and India are just a few places that followed suit. Who knew what a small beach town with an iconic name could do? This ban was not the solution to the entire plastic pollution crisis, but it is a key moment because it is a tangible step forward that opened the door to much more change. At that time many large-scale organizations were trying to think bigger and convince lawmakers to make more comprehensive bans on single-use plastic, and they are still trying today. I often compare solving the plastic problem to eating a hamburger in bites versus trying to shove the entire thing in your mouth all at once. I wholeheartedly believe that the attempts to tackle the entire plastic pollution crisis at once is a strategy that does not work because it's too big of a bite to take on and implement successfully. This is about a transition to break free from our dependency on single-use plastic that needs to be well thought out. In order to be successful, nonplastic alternatives must also be available for businesses to switch. For example, plastic lids for coffee cups were not included in the Malibu ban because there weren't enough alternatives available for businesses to conduct their operations successfully. Creating change in a small town with a big name like Malibu was a major win that pushed the needle forward while making major waves around the world. This is where the idea of the "imperfect environmentalist" was born. You see, had we gone to the City of Los Angeles and demanded a ban on all single-use plastic, then we would have failed and perhaps still been at square one. However, the approach we used—to go to a smaller town—allowed us to create the momentum needed to begin the bigger effort (which we are still in) to combat plastic pollution as a whole. We don't want perfection to be the enemy of good enough, so we have to start somewhere. All of us.

The day the ordinance was put into effect was also a very memorable day. For the first time in history, McDonald's in Malibu provided paper straws, and Starbucks was forced to provide paper straws as an alternative for their unfortunate yet infamous green plastic straws. I was among the first people to witness this, along with Senator Henry Stern. We have video of both of us doing a little happy dance because we had just forced sustainability within these mammoth corporations. This would never have been a priority for them had it not been for the Malibu ordinance.

The Malibu ordinance was made public so that everyone could read it, learn from it, and adopt it in their cities. The domino effect began, and it was bigger than we ever imagined possible. We saw city after city, state after state, and country after country create their own bans on single-use plastic straws. To this day, many use language from the ordinance created by the City of Malibu, and I am still so proud to be a part of this historic ban. Change is possible anywhere, but it all depends on persistence, comradery, patience, and, of course, how you present your case. I never shamed or blamed anyone. Rather, I invited people to think differently and to lead the way with new ideas and solutions. People want to be a part of that energy as opposed to having someone

Figure 2.4. Receiving a drink with a paper straw at Starbucks after the ban on plastic straws in Malibu. Photograph by Sheila Morovati.

try to strong-arm them into making a change because they are doing something "bad." In the next few chapters you will see that this positive tone is consistent.

One supporter approached me at an event and shared that she felt my work was like that of an airline marshal (the person on the runway who guides the plane to its final destination). She said that I use the marshaling wands to ever so gently yet effectively veer people to see another way. I think about that comment a lot and realize that I have a lot of faith in humanity and believe strongly that people do not inherently want to harm this planet. However, with the way society forms and changes over time, we become accustomed to new ways that are not positive for the health of our planet. My goal is to step in and shine light on those new habits that have formed, and reset them to more sustainable ways. If I am able to do this with my "marshaling wands," then that is a huge win.

One unusual result from this success in Malibu was that it left many people wondering what a crayon organization had to do with a ban on single-use plastic. I understood the confusion, but I kept explaining to everyone that the straws and the crayons are both just a "habit of waste" that we partake in but don't even see anymore. Even my board members expressed how their network of friends also didn't see the connection between straws and crayons. They questioned why we were working on such separate issues within the same charity. A very honest board meeting was all it took to push me to do something new. The night after the board meeting it occurred to me that Crayon Collection's 501(c)(3) nonprofit standing is based on environmental efforts and that we could create another program with its own website, logo, and so forth. In a big "aha moment" I decided to name it HOW, short for Habits of Waste. This is the perfect name because, with my sociological approach, I was eager to shine light on all those habitual behaviors we encounter each day that pollute our planet so greatly but that we often didn't even notice anymore because they have become so routine. I wanted to show people different options and offer solutions that didn't take much effort or energy so that the masses could be a part of the movement too. I wanted large-scale society to become "imperfect environmentalists" instead of assuming no responsibility since they weren't technically "environmentalists." My goal was to prove that if you are a living, breathing human being existing on planet Earth, then you are automatically an environmentalist and you *can* make a difference every single day no matter how big or how small the action is.

There was a great energy and flow behind Habits of Waste. I was able to quickly secure the URL, create a logo, and build the website within a few short months. I will always be grateful to a young woman working for me at the time named Julia Greene, who was my right hand and helped me develop all the assets needed for the website to go live. I realize now that it probably would have taken me double, possibly triple, the amount of time had it not been for Julia.

Within three to four quick and very productive months www.habitsofwaste.org was ready, so we threw a super elegant launch party at my friend Beate's ultramodern home in Brentwood, California. It was a beautiful event that filled me with so much more hope. The truth is that Crayon Collection, my first baby, had developed into a very well-oiled machine with its wheels in motion. My team and I continued to work hard on Crayon Collection, but I was excited to take on even more. I wanted to know what else I could do because one thing was for sure: I was quite capable of convincing people in the most gentle and inclusive way to change their habits. I owe a lot of this learned quality to one of my favorite books, written by Dale Carnegie, *How to Win Friends and Influence People.* Carnegie figured out how to get people engaged, interested, and open to ideas without any force or any strong-arming. He also adds that most people are generally the same in the bigger picture, and understanding the nuances of the human ego is a key part to making change.[8] I highly recommend this book to everyone in any industry because it touches on human connection regardless of what type of end goal you have in mind.

The interesting thing about Habits of Waste is that we really didn't have a clear understanding of what we were going to do as an organization. Were we educating the public? Were we activists to protect the planet? What were we? This is where I often tell people to take just one step at a time and the rest will come. At that point all I knew for certain was that I needed a website and a logo, which became my first step. My priority was to build a place for people to learn about us, and to create a solid brand. The rest would come after those essentials were in place.

Once www.habitsofwaste.org was up and running, I decided that I would begin by tackling the most common issues that kept coming up in my conversations with people around me. You see, many "whataboutisms" popped up quickly after the plastic straw and cutlery ban in Malibu. The definition of a "whataboutism" is a behavior phenomenon that people continue to do this very day where they respond to an issue by raising a different issue.[9] The press does this often. During the press blitz after the Malibu ordinance passed, I was interviewed by several journalists. Many of them would usually start with a cordial compliment of my work thus far, and then the "but" would come. They would say, "But what about plastic water bottles?" and, "What about all the plastic cutlery I get when I order in?" or, "What about plastic cups?" To each one I would answer patiently that I was happy to hear that people were thinking bigger and perhaps the plastic straw ban was low-hanging fruit that could be used as a stepping stone for more lofty environmental goals. I decided the role of HOW would be to tackle each of the "what-aboutisms" at the appropriate time.

Determining the right time to address each of the new environmental challenges people assigned to me was always challenging. There were so many issues to resolve, but I stayed true to my instinct, which was to conquer the ones that annoyed most people. For example, the "what about plastic cutlery?" issue seemed to come up often. People were frustrated over being constantly bombarded with plastic cutlery even if they didn't need or ask for it. This concern was shared with me as though it were on repeat. So many people shared the same concern and felt frustrated that single-use plastic (particularly cutlery) had crept into our everyday lives. Perhaps the Malibu plastic straw and cutlery ban gave people a sense of hope that even a mammoth problem like single-use plastic cutlery could be solved.

Habits of Waste's first campaign was #CutOutCutlery, which put us on the map for a second time.

HOW YOU CAN BE AN IMPERFECT ENVIRONMENTALIST

1. You can try to have plastic straws or plastic cutlery banned in your city too. Most cities in the United States gives their community members two to three minutes to speak at city council meetings. This is a huge opportunity to share your concerns with your city council members. Learn when your next city council meeting will be held by visiting the city website or calling city hall.
2. Speak out and state your case without blaming or shaming anyone. Speak in a clear and matter-of-fact way. State the problem and then offer your solution immediately thereafter. We all know that there are so many problems, but if there is a solution to be had, then the chances of making change are exponentially greater.

3. Become friendly with city administrators. They hold the keys to many things. You can do this by finding their names and emails on the city website. Reach out to them, introduce yourself, and ask to learn a little bit about their current goals. Check in on them every so often and say hello the next time you see them in person. (I will always be grateful to Casey Zweig and the in the City of Malibu administration for taking the issue so seriously and creating the gold standard for single-use plastic bans.)
4. Go to a city council meeting to familiarize yourself with the format and gain an understanding of how the system works. Then you might be more comfortable to go to the next meeting and use your two to three minutes to bring up something that is on your mind.
5. Refuse plastic straws whenever possible. We are often bombarded with single-use plastic, so speak out when you order a drink and tell the waiter, "No straw please."
6. Are you a straw lover? No problem—go reusable! Bring your own reusable straw(s) when you leave home.
7. Compostables are not the answer. They behave just like plastic in the environment unless they are properly composted, and they cost more.
8. Watch the documentary *Straws*. It's a great tool to help everyone understand the issue with single-use plastic.
9. Share the *Straws* documentary with your local schools so they can also screen it for their students.
10. Let the restaurants and cafés you frequent know that there are better alternatives to single-use plastic straws that they can provide to customers upon request, especially since most people don't even take a straw if it's not already in their drink. This way, they can offset the cost difference of pricier eco-friendly alternatives.

3

#CUTOUTCUTLERY

In the study of sociology a plethora of research is done on default settings. Did you know that most people will never change even one default setting when they receive a computer from the manufacturer?[1] This was a great piece of information to have as I started to tackle the recurring complaint about the excessive amount of single-use plastic cutlery people are bombarded with when ordering in, followed by how guilty everyone feels for throwing the unwanted and unused plastic cutlery away. Many avoid the guilt by creating a junk drawer to hold these plastic utensils for ages until there is no more space, which is when they'd guiltily discard the unused packs of plastic cutlery. Since this is such a common issue, I decided to take it on as my first campaign under Habits of Waste. Plus I felt that plastic cutlery was a great issue to address since the Malibu ban included plastic straws *and* plastic cutlery, so it seemed like a natural progression.

For most people, eating takeout at home means they have access to stainless steel cutlery, which most people prefer, so a large portion of single-use plastic cutlery included in food delivery orders is never even used. Not only is it bad for our planet, but it's also hard on restaurants' bottom line because, after all, they are spending their hard-earned dollars on these single-use utensils as a convenience for their customers, yet most aren't even wanted. The problem had been getting worse every day, as the number of people ordering in has soared over the last decade. Ordering in was on the rise thanks to all the new and convenient food delivery applications available for free download coupled with the stay-at-home restrictions from the COVID-19 pandemic.[2]

My friends, family, colleagues, supporters, and even social media followers had one thing in common: They all wanted to do better by refusing the plastic, yet they felt powerless because no matter how many ways they tried to communicate with the restaurant through the food delivery app, they somehow always received the unwanted plastic cutlery with their order. Many of my environmentalist friends called it the "nonconsensual" receipt of single-use plastic. In food delivery applications there is usually a box to add a note to the restaurant at the end of the food order placement process. However, no matter how many times I wrote a note to state my "NO plastic cutlery" request, I would peek into the bag and be disappointed to find yet another stack of plastic cutlery.

Before I continue, I would like to make one clarification because I can just imagine how many readers are saying, "Why can't we just recycle the plastic cutlery?" Well, because single-use plastic cutlery is not recyclable, so let's start by getting that out of the way.[3] The shape of plastic utensils makes them too small for recycling machinery, and the variety of plastics they are made of are virtually indistinguishable, leaving plastic utensils nearly impossible to sort properly for recycling. In fact, only 9 percent (if that) of all single-use plastic gets recycled.[4] The rest ends up in landfills, rivers, or our oceans, harming the environment and the precious creatures that inhabit it. This isn't an easy problem to solve. The numbers are astronomical! In the United States alone forty billion pieces of plastic cutlery are produced per year with the sole intention of being used once and thrown away.[5] Sometimes they are not even used but still thrown away. I knew there had to be a solution, yet inherently I realized that convincing everyone in the world who ordered takeout to make the conscious effort to comment and request no plastic cutlery was impossible. I did know that there had to be a solution to reduce the demand for these pesky plastic utensils. After all, if most people didn't want them, why did so many people still receive them?

The default setting was the answer. I decided to create the #CutOutCutlery campaign as a call to action for Uber Eats, Postmates, Grubhub, and DoorDash to change their default settings so that no one receives plastic cutlery unless they request it by either sliding the toggle or clicking the "+" to ask for a certain quantity of plastic utensils when ordering through apps or restaurant websites. This would surely solve the problem, as it would create a blanket change that allowed the smaller population of people who wanted plastic cutlery to request it, while the rest would automatically not receive it. This way we could protect the planet from unnecessary trash *and* help restaurants reduce the money spent on their bottom line. (It's a common misconception that plastic cutlery is free, but someone is actually paying for it as part of the restaurants' expenses.)

The next challenge was how to reach these massive food delivery applications so that they would listen to us. They are the go-between for millions of restaurants worldwide. Uber Eats alone has nine hundred thousand restaurant partners![6] I thought if I could just convince the big four apps to comply with the change, then we would be effecting change globally. Once again, it was low-hanging fruit that would be a win for everyone. Looking back, I realize I had a small sense of naivety, which really helped me. As per usual I just put one foot in front of the other, and sure enough, the right doors opened for me. We are a relatively tiny organization with a small social media account, which I knew wasn't going to provide the leverage we needed. However, there was a *huge* number of people who were extremely annoyed with the plastic cutlery issue.

As I said earlier, talking to other people about your ideas and work is key. I really admired an organization around at the time called SameSide.[7] This group created events and activated people by reaching out to their congresspeople, mayors, and other government representatives by text or email. I loved how fresh and young their approach was and how tools like text and email were available to help create change. What better way than to get young people on board? Have a party and get them activated on their phones!

I had a mutual friend who regularly went to SameSide events. I reached out and asked her if I could have a meeting with the founder. She obliged and my friendship with Nicole àBeckett began. I explained to Nicole that I loved her fresh and innovative approach to getting new laws passed and wondered how we could use a similar approach toward environmental issues. She explained that she was using what was called a P2A, or "phone to action," platform, which could be tailored to reach government entities or even large corporations. In a nutshell, it was a conduit to reach anyone we wanted, with the help of the masses. For example, we would write a convincing email and find the contacts and the corporations we were targeting, but the benefit was that anyone could send that email using the platform simply by signing in. The recipient (in this case, the food delivery applications) would receive the same email from thousands of unique users with unique emails. It allowed us to authentically express the concerns of the public by giving individuals the power to send their own emails to the entities we were trying to create change within.

The problem with launching our own P2A campaign was the cost. Once again, sharing my story and inspiring those around me was key. Nicole explained that she could create a campaign for us under the SameSide umbrella of campaigns for a nominal monthly fee to help her cover some of the costs. This was the key piece to get #CutOutCutlery going. The truth is that budgetary constraints are a serious reality for startup nonprofits, but luckily we had several generous donors who helped us get on our feet. The seed money we raised provided a small amount of funding to help pay for the overhead costs we had incurred. Needless to say, had Nicole not presented us with such a generous offer, we probably could not have done what I am about to share with you.

My next step was to write a strong email to convince Uber Eats, DoorDash, Grubhub, and Postmates to change their default setting for plastic cutlery to be opt-in via a toggle so that users no longer receive plastic cutlery *unless* they request it. Next, I had to find email addresses to send the request to. After thinking about it long and hard I realized that sending the email to customer service would be the right thing to do. Here's why: I knew that this P2A platform allowed people to send the prewritten email using their own email addresses, so it was very different from a petition, and I expected thousands of emails to be sent in. For this reason, I didn't target any individuals at these companies since they would easily be able to change their email addresses, plus I never want my organization to feel like we are harassing or singling an individual out. Another benefit of sending our letter to the customer service inbox was that they received the same prewritten email thousands of times from thousands of different people, and after a while it would only be natural that someone in customer service would take notice and send it up the ladder to the right department. To me this was a huge plus over a petition because it enabled us to effectively and efficiently infiltrate these companies.

Below is the original email I wrote to send to DoorDash via the SameSide P2A platform. This email was replicated for all the food delivery applications, including Grubhub, Uber Eats, and Postmates, and sent to their respective customer service departments.

> Today, I am joining Habits of Waste (HOW) to ask DoorDash to change the functionality on its app so that consumers can refuse plastic cutlery. We also are working with state lawmakers in California to turn this practice into law. DoorDash can simply change the default setting on their application so that plastic cutlery is only provided upon request.
>
> Sadly, in the U.S. we are discarding 40,000,000,000 pieces of plastic cutlery per year, which never decompose and end up in our precious oceans.
>
> Many DoorDash customers order meals to eat at home and don't need or want plastic cutlery. Restaurants would happily reduce costs by customers receiving cutlery by request only. It would be a simple functionality change where the default setting automatically does not include plastic cutlery yet, those who need plastic cutlery can *opt in* and receive it with their order. Many DoorDash participating restaurants have been interviewed and are open to this change.
>
> Plastic does not biodegrade; it breaks down into smaller and smaller toxic micro particles, which contaminate and bleed into the environment. All this single-use plastic ends up being digested by our oceans, marine life, ultimately ending up in our food and water.
>
> DoorDash can help change this "Habit of Waste."
>
> As a HOW Changer, I hereby request that DoorDash change the default setting on their application so that plastic cutlery is only provided upon request.
>
> Thank you for your time and consideration. This simple switch will be revolutionary just like your service. We encourage DoorDash to join in and help lead the way for other companies and corporations—the environment needs you to pave the way to heal the planet. Please visit @howchangers and www.habitsofwaste.org for more information.

In a simultaneous effort, I found the head of sustainability for each and every one of these food delivery applications and began my personal outreach to share the campaign details and give them a respectful heads-up that these emails were coming their way. It is the head of sustainability's job to consider the company's environmental impact and implement sustainable practices. Therefore, I wanted them to know why we were doing this and that we had a simple solution to really fix a huge problem. From the outside I felt the request we were making wasn't *that* big of a deal, but in actuality it was a huge issue. It's like we were coming in and asking them to rearrange the furniture in their living room. Every single element in these applications has been tested and studied and analyzed and reanalyzed. It is very challenging to make any changes without more testing, studying, and analysis. I recognized that it really was a massive lift for them, but with enough public pressure I knew we could make the change.

Uber Eats came first. Our steady stream of emails requesting the default setting change, plus a lucky personal introduction to the sustainability department at Uber Eats, gave me a chance to explain the benefits of this default setting change, which would both promote positive change

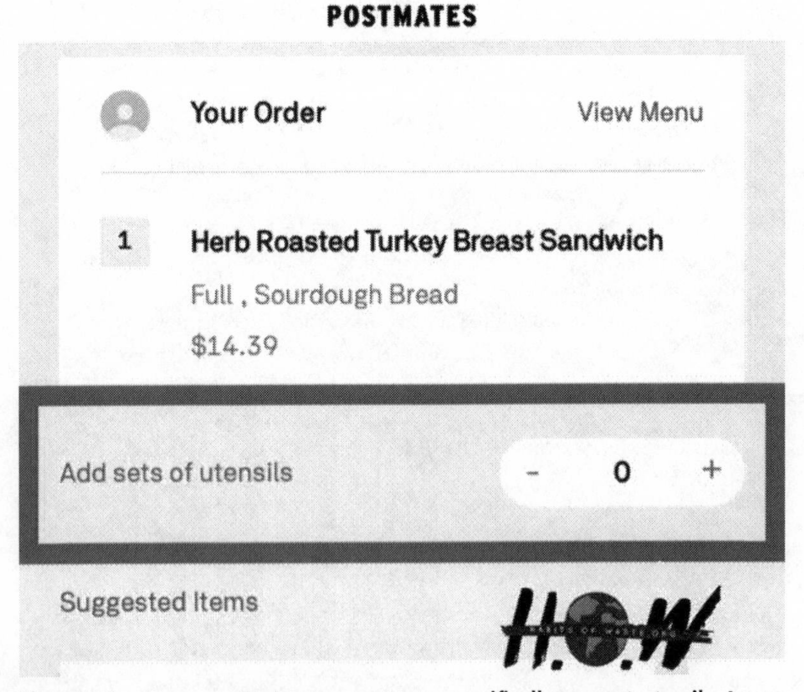

Figure 3.1. Users on Postmates must now specifically request utensils. Image courtesy of Habits of Waste.

for the environment and provide significant cost savings for restaurants. As expected, the sustainability team was all for it, but any outward-facing changes that affect user experience are seldom made, so that was the next layer to face. After enough conversations and with a willingness from all parties to do better, I was invited to work closely with the team at Uber Eats to help them with this update. We tried to use language that would inspire people to know that they were helping reduce waste and protect the planet right then and there just by using the app. I had to sign a nondisclosure agreement to work with them on this process, which I still respect, so I am unable to share any specific details of my time with the engineers at Uber Eats.

Once I had the interest of the Uber Eats team, things went very, very fast. We had multiple Zoom calls with their sustainability team, with people logging in from all over the world. I remember one of these calls had to take place while I was on vacation with my family in Lake Tahoe, so I dropped everything to work for several hours on the #CutOutCutlery × Uber Eats movement, as opportunities like those are few and far between. As best practice, if there is ever a company or individual you aim to meet with, please accept any or their suggested call times unless you have a major scheduling conflict. The simple effort of finding a mutually agreed-upon time could cost you the whole meeting. I believe we must try our hardest to make the connection as easy and positive as possible.

Uber Eats was in! I was thrilled, to say the least. We got Uber Eats (Uber Eats!) on board. What a huge win that was! I just could not believe it. At that point we had sent about fifteen hundred emails to them from the P2A platform, which I knew had made an impact because the sustainability department's first request was, "Now can you please stop sending us those emails?" I gladly obliged and took Uber Eats off the emails that were going out.

The first step for implementation was that Uber Eats had to pilot, or test, the default setting change in a few major cities to see how customers responded. Luckily it was a nonissue, and no one had any complaints. After a positive pilot program in a few key cities, Uber Eats launched #CutOutCutlery within the state of California, then across the entire United States, and globally soon thereafter. This shift was quick and effective because a pillar of Uber Eats and all the remaining food delivery applications is that every user is meant to have the same experience with their app no matter where in the world they are located. The consistent experience is of utmost importance, so this was a gift for my effort. I was so glad I didn't have to convince the apps to make the change in different cities, states, countries, and so forth.

AND THEN THERE WERE THREE

I knew that until all four major food delivery apps joined in, we had the potential for massive confusion. You see, restaurants usually work with multiple food delivery apps and try to fulfill orders quickly so the drivers can pick up and deliver them in a timely manner. To say a restaurant

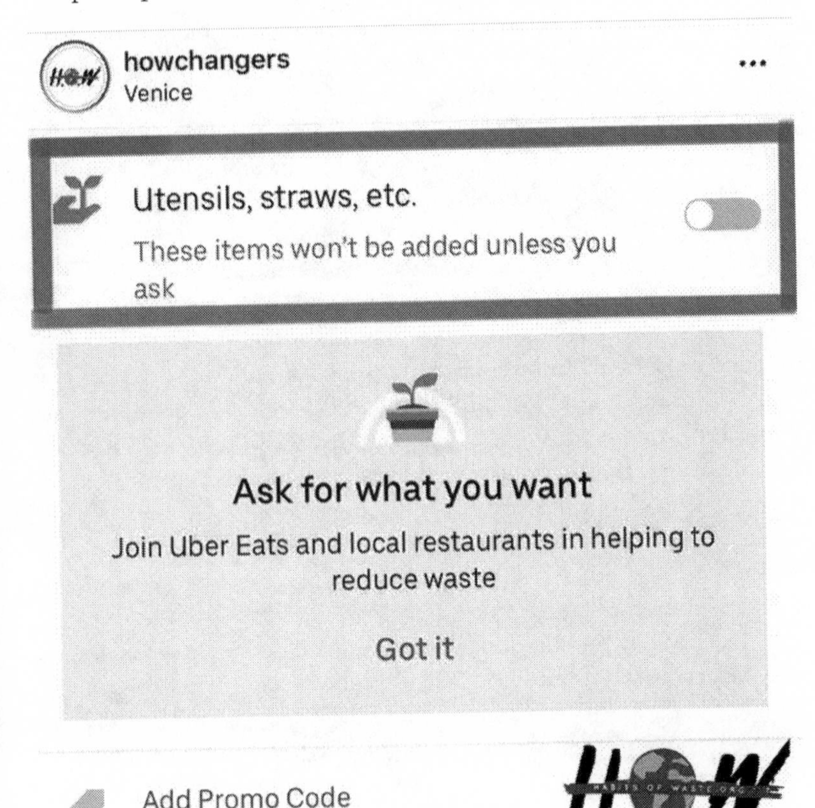

Figure 3.2. A flyer for the #CutOutCutlery campaign by Habits of Waste. Image courtesy of Habits of Waste.

Figure 3.3. In-app changes on Uber Eats thanks to the #CutOutCutlery campaign. Image courtesy of Habits of Waste.

kitchen is hectic is an understatement, so we really were looking to create a new norm that wouldn't require any thought while filling the meal orders. This meant we really needed all apps to share the same default setting around plastic cutlery.

Luckily, soon after Uber Eats agreed, we were able to convince Postmates, which was the smallest app in the bunch. They were amazing to work with, as their sustainability team truly cared, pushing for the opt-in function for plastic cutlery and making it happen with lightning speed. The leadership of Uber Eats adopting #CutOutCutlery was a huge factor in Postmates' quick action.

As time went on we wanted to know the impact of this opt-in/opt-out functionality. Postmates shared their internal data with us, showing that within one year of joining the #CutOutCutlery campaign they saved 122 million packs of plastic cutlery from entering the waste stream. I took that number and multiplied it by the cost of one pack of plastic cutlery to find that the total restaurant savings was a whopping $3.2 million. The impact was huge, and these numbers allowed us to blaze forward.

The two remaining apps, Grubhub and DoorDash, were also among the largest ones. I knew that I couldn't solely depend on my tiny but mighty supporter base to send emails, so I reached out to BuzzFeed to help us raise awareness. Thanks to Auri Jackson, who was the environmental video producer at that time, we were able to create one of the funniest and most impactful environmentally focused videos of all time.[8] We had already established a great relationship with BuzzFeed following a story Auri did on my work after the plastic straw ban in Malibu, so she was willing to create this #CutOutCutlery video too.

The new #CutOutCutlery video begins with two very familiar faces, actors Adam Devine and Laura Dern, among several others, including Senator Ben Allen. The idea was to open the video with something silly that makes people lean in and wonder what we were talking about. While the script actually references the junk drawer filled with plastic cutlery, we disguised it with lines like, "We all have one," "Even your grandma has one," "If you show me yours, I'll show you mine," and "It's my dirty little secret." The camera then cuts to Adam Devine opening the plastic cutlery drawer in his kitchen as he makes a sour face. After the opening segment, we go into information about the #Cut-

OutCutlery campaign with the aforementioned stat from Postmates, which made everyone take the campaign very seriously.

We did this because I strongly believe that the doom and gloom of environmental public service announcements does not inspire people to take action. I think a sense of fear has been instilled among the public, and without specific tasks to join in on, people freeze because they feel too small to do anything; so, in turn, they do nothing. We felt that if we captured peoples' attention with something fun and funny, they were more likely to watch the video and get involved. We were right. The video was shared across all of BuzzFeed's various platforms to a total of 650 million followers. We increased our emails to the food delivery apps tenfold thanks to BuzzFeed.

I was simultaneously reaching out to anyone I could at Grubhub and DoorDash, yet I was hitting a wall. No one would take the effort seriously, and under no circumstance would anyone take my calls or answer my emails. Usually with patience and persistence I would find a way in, but this time I felt totally blocked from all communication with the sustainability and marketing teams. I even contacted Assembly Member Ian Calderon, who was responsible for passing AB 1884 in California, which made plastic straws available only upon request. We met several times and thought he could help enforce the same concept, only this time for plastic cutlery. After many conversations with his chief of staff, I was told just before the legislative session that he was retiring to spend more time with his young family. One curious factor during those calls was that he kept asking me what other organizations were in support of this movement. Did I have groups like the Sierra Club willing to make a statement? Who else was involved and believed in this campaign? I realized I needed more people behind me.

I decided that it would be helpful to share the #CutOutCutlery effort with a large coalition of plastic-pollution-fighting organizations. I met with several individuals who were closely involved with the coalitions, and they invited me to share the work I was doing with over one hundred environmentally focused participants nationwide. We held the meeting via Zoom, and overnight huge organizations like Oceana, Upstream, 5Gyres, and many others began to support us. It was a huge hit and I was thrilled to see how they all got involved and worked with us to continue sending emails via our P2A platform. Eventually, after fourteen thousand emails were sent, we finally convinced Grubhub and DoorDash to join in as well. Both made the changes globally, so we were now well on our way for a new societal norm to be born whereby people no longer expected plastic cutlery with their take-out orders. Each of these food delivery applications helped tremendously by explaining to their millions of users that plastic cutlery would be available only by request. This interaction with so many people started to open the mindset of wasting less just by ordering in!

The coalition suggested we change the name of the campaign to #SkipTheStuff so that it could be broadened to include condiments as well. Together we decided that #CutOutCutlery should continue as the public-facing name since the campaign had already made so much traction and was recognized by food delivery apps and the public. But we decided to use the broader #SkipTheStuff name for legislative efforts.

With this win, the "plastic fighters" coalition decided we should work together to create legislation in the City of Los Angeles and possibility the State of California to make the opt-in functionality for plastic cutlery required both in food delivery apps and within restaurants. At this point it was 2021 and we were still in the middle of the COVID-19 pandemic. No one would pass a law to put more pressure on restaurants, so my calculation showing the monetary savings for restaurants was a major element to creating a new law. The idea of having a legislative measure was helpful because we would never have to convince another app to join in, nor could they change their default setting back to what it was before. The City of Los Angeles adopted the #CutOutCutlery/#SkiptheStuff campaign first. After that major win we set our sights on Sacramento.

Assembly Member Wendy Carrillo presented the campaign to the California State Senate, which voted yes and then passed the law statewide (AB 1276). Part of the arsenal we used to prove our case was the sassy BuzzFeed video done on the #CutOutCutlery campaign, which quoted numbers from Postmates that I had calculated into the dollar amount of $3.2 million in savings after just one year of adoption. The passing of AB 1276 meant that every restaurant *and* food delivery app in the state of California could no longer provide plastic cutlery to customers except by request. Since then, the State of Washington has passed the very same bill and most recently New York City adopted #CutOutCutlery/#SkiptheStuff. We expect to see many more states join in.

The campaign is still active today. We are now targeting and emailing the major fast-food chains McDonald's, Burger King, and Wendy's to provide plastic cutlery upon request globally. We just removed Chipotle from the list since they made the change we requested on their app and in all their restaurants, so we are celebrating! Once we get the rest of the fast-food chains on board, we will have achieved a new societal norm making plastic cutlery something that is no longer offered without a request. You join in too! Send in your email today by visiting the #CutOutCutlery campaign page at www.habitsofwaste.org. The good news is that McDonald's in the United Kingdom and Australia has already joined in. France banned plastic cutlery altogether (among other single-use plastic items) as of January 1, 2022. India's ban on plastic cutlery and other single-use plastic items took effect on July 1, 2022. The United Kingdom followed with a similar effort that went into effect on January 14, 2023.[9] More and more countries are following suit. This quick success is partly because 1.96 billion people interface with food delivery apps every day. This scale allows for new norms to infiltrate society much faster than ever before.[10]

Needless to say, we have reduced the demand for plastic cutlery, but we have a long way to go. We are so proud to have created a new system that allows everyone, everywhere, to do a little better. Often I envision a busy family that is just trying to get dinner on the table and make it through the day. I am relieved by the notion that this family doesn't have to think about plastic cutlery at all thanks to #CutOutCutlery. Mission accomplished.

HOW YOU CAN BE AN IMPERFECT ENVIRONMENTALIST

1. Consider refusing plastic cutlery if you don't need it when you get takeout. If we all repeatedly state our request for less trash, then eventually restaurants will stop ordering so much of it and we'll reduce our impact on the planet.
2. Bring your own! Simple actions like having a set of steel cutlery in your car or asking your office to have reusable cutlery in the kitchen is all it takes. Food psychology studies show that people

enjoy eating with stainless steel cutlery much more, so this is an easy switch!

3. Start a campaign like this in your community. One of my regular sayings is to dream big and imagine throwing spaghetti at the wall. Sometimes it sticks and it's a "yes," and sometimes it doesn't. But if you don't try, you will always have a "no." My advice is to just go out there and put one foot in front of the other to begin taking the necessary steps toward your goal.
4. Create change by offering solutions to problems that are beneficial to everyone. Complaining without offering a solution won't get you anywhere.
5. Start sharing your sustainable ideas with friends and colleagues. You might be surprised at the amazing things that happen, such as doors opening or connections being made with incredible people who help you make it through. Nothing I have ever achieved was from my efforts alone. It has taken many people along the way who believed in this work and helped push it along.
6. Be positive and don't give up. If you hear many people say "no" to you, just know that you are getting that much closer to a "yes."
7. If you didn't ask for plastic cutlery and it ends up in your order anyway, please take one minute to call the restaurant and gently remind them that you did not request it. Then they can look at incoming orders more carefully next time.
8. Download our #CutOutCutlery signage and print out a few copies. Share these with your local restaurants. They will appreciate the clear messaging for their customers to understand that they will need to ask for plastic cutlery moving forward. It's a huge money saver for restaurants, so they are very happy to join in.
9. Sign on to our P2A! We are still targeting major fast-food chains globally, so the more emails we send out the better.
10. Share to social media. Perhaps share a picture of your lunch that included reusable cutlery or tell your followers that you just joined the #CutOUtCutlery movement so they can join in too.

4

#LAGREENTEEN

This was probably the most eye-opening and heartbreaking period of my work. It was the first time red tape bureaucracy really had me feeling stumped, defeated, and dumbfounded, but somehow, despite all odds, we still made a huge amount of noise and ultimately created the change that was desperately needed.

I remember sitting in our very first office space, which was a shared workspace near Venice Beach in Southern California. I was at a point where I realized the organization was growing and I needed to raise more funds. I had so many ideas and wanted to grow so badly, so I decided to contact my bank to see if there were any grant opportunities for a small, growing nonprofit. It was 2018, the plastic straw ban in Malibu was well underway, and Habits of Waste was just beginning to take form.

As I was looking for ways to increase our impact and get funded, I connected with Wells Fargo bank to explain my idea of visiting our lower-income inner-city schools to help the students understand more about climate change, the planet, and the role they could play to help protect it. This was a huge priority for me because studies show that vulnerable communities are less likely to act on environmental initiatives.[1] Part of the reason for its this could be the geographical disconnect from nature. For example, although many of the inner-city schools in Los Angeles are situated about seven miles from the beach, some had never made the trip to visit the ocean. Part of my idea was to create a connection with the ocean for them to share ideas, providing the courage to take action to help.

After an extensive grant application and many phone calls, we were awarded a $10,000 gift from Wells Fargo to fulfill this idea to teach inner-city students about environmental issues over the course of four weeks. The plan was that each Friday we would bring a different guest speaker and cover topics about the environment to a group of students as an added layer to their science curriculum. It was clear that we needed to teach environmental justice as a portion of this programming. Per the U.S. Environmental Protection Agency (EPA), "Environmental justice is the fair treatment and meaningful involvement of all people regardless of race, color, national origin, or income, with respect to the development, implementation, and enforcement of environmental laws, regulations, and policies." However, here is one basic example of the discrepancies: "In a two-year long study, a team of researchers from the Nature Conservancy found that 92 percent of low-income blocks in the U.S. have less tree cover and hotter average temperatures than high-income blocks. The inequality is most rampant in the Northeast, with some low-income blocks in urban areas having 30 percent less tree cover and average temperatures 4 degrees Celsius higher than high-income blocks."[2]

Implementing this program was also important because Black, Brown, and Indigenous peoples continue to be excluded from environmental policy, conservation, and public health issues.[3] Despite this, these communities face the biggest burden of climate change. The EPA released an analysis showing that "the most severe harms from climate change fall disproportionately upon underserved communities who are least able to prepare for, and recover from, heat waves, poor air quality, flooding, and other impacts."[4]

To make matters more complicated, the community we were going to teach was the target of major marketing campaigns for the largest polluters in the world, Coca-Cola and PepsiCo. Research finds that these companies target low-income communities, with a positive relationship being found, for example, between the level of soda marketing and Supplemental Nutrition Assistance Program (SNAP) benefit issuance days.[5] This serves to increase diet disparities across communities of different incomes and demonstrates an exploitation of the poor by large corporations.

We selected Los Angeles High School to pilot the new #LA-GREENTEEN program in. The school ranks in the bottom 16 percent of the Los Angeles Unified School District (LAUSD) and is considered Title 1, which means federal funds help the school address students' educational needs and enable them to meet academic standards, as well as close the achievement gap. The student body is composed primarily of Black and Hispanic students (roughly 90 percent), followed by Asian (10 percent) and White students (0.8 percent).

The four-week structure was meant to inform the kids about the climate change and environmental stewardship, empower them to understand what environmental justice was and determine whether they had faced it (they had), and then provide them with steps to take action, including how to create a social movement via social media and effectively use their voices to make change. I personally wanted to help them amplify something that was important to their community with tangible steps that Habits of Waste could continue to support long term. The name of the program was #GREENTEEN and it was meant to be scalable so that we could add the city name of each area at the beginning. My wish was that after #LAGREENTEEN we could start #CHICAGOGREENTEEN or #MEMPHISGREENTEEN, for example.

As a part of the curriculum, we brought in four different speakers (one per week), including a UCLA professor who talked a lot about social justice and environmental justice, and then I spoke about Habits of Waste and how each person has the power to create change. A very important speaker we contacted was Dr. Tim Pershing, a senior field representative from Representative Richard Bloom's office who spoke on the third Friday and shared the problems our planet was facing. (Having his participation was a major key factor to the outcome we achieved, which I explain later in this chapter.) Finally, the social media experts from ATTN: media came in to teach the students how to take action through social media.

In preparation for this program, I met with the educators and administration at Los Angeles High School. I shared details of my plans for the students and excitedly announced that each student would receive a reusable water bottle, which would be provided by a sponsor at no cost, so that the students could refill their new water bottles and end their dependency on plastic water bottles.

What happened next shocked me to the core. The administration looked at me earnestly and said, "Where do you expect the kids to refill their water bottles?"

I announced naively, "At the water fountain, of course." Well, that was a silly assumption. Apparently, they would have no use for reusable water bottles because there would be no place for them to refill them. The water fountains were something from another era. They were dilapidated and filthy, not to mention the questionable water quality. They did clarify that in an emergency situation the students could let the water run for one minute before having a sip in an effort to flush out all the lead. This was upon emergency only. Did they really just say that to me?

I did a walk-through of the school and was stunned. Most of the water fountains were covered with brown plastic bags, indicating they were out of order. Or they were covered in black soot and had not been cleaned for years. How could anyone drink out of those? Was I really in one of the wealthiest cities in the world—in the most powerful country in the world? Was I really in the second-largest school district in the United States? This was absolutely unacceptable, and I could not in good conscience sleep at night knowing these students did not have access to clean water. The students needed new water hydration stations. That became the mission of my four weeks with them.

Figure 4.1. Water fountains in a state of disrepair at Los Angeles High School. Photograph by Sheila Morovati.

One of the first phone calls I made was to the LAUSD communications department requesting permission to document the current situation and the changes we were going to make with students. I explained to the person in charge that we wanted to shine a light on an issue within the school district for the leadership of LAUSD and Los Angeles High School so that other schools would follow suit and create the change in their schools as well. I was happy to receive a thumbs-up and immediately invited BuzzFeed to document the work we were going to do.

The environmental video producer from BuzzFeed came through the school and took some footage of the water fountains, and to this day I remember her voice in my mind because she too was stunned. It was horrific. Then she began to interview the kids. She asked the students how they stayed hydrated. Most said they drank milk because it was free at school. Many brought their water from home, but most ran out by lunchtime. Some had headaches because of dehydration, and paying $1.50 daily for one Dasani water bottle was just too expensive, especially since $1.50 was also the price of their bus ride home. The student athletes had it the worst because they had no water during practices unless they drank questionable water from the fountains. The school explained that they used the funds from the plastic water bottle sales for a variety of things, which is why they still needed to charge the students for those water bottles.

What baffled me was that many schools in LAUSD already had new hydration stations. I was so curious to understand why and how they had them, but Los Angeles High School didn't. Well, the answer was simple. The schools I had seen were in well-served areas, so the parent body raised funds to install the new hydration stations and continued to make regular upgrades to the school. This was one of my first exposures to environmental and social injustice, and I was deeply distressed by it.

If it was a matter of raising money, then I knew we could easily do that. All we had to do was post the BuzzFeed video and show the world what was happening, and we could swap out the old fountains for new ones. When we began our fundraiser for the new hydration stations, I reached out to a new sustainable water bottle company, which was willing to donate seven Elkay refill stations. The cost of installation was meant to be nominal so the money we were raising could go toward the installation instead of the actual new water hydration stations.

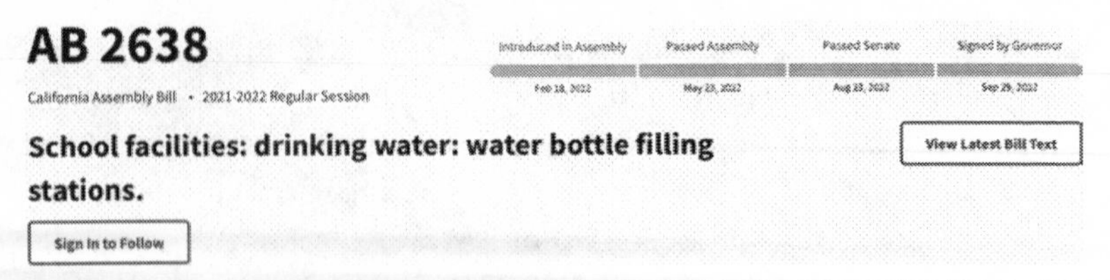

Figure 4.2. Bill AB 2638 requiring schools to provide free, fresh drinking water for students.

I was thrilled that we had so easily resolved the problem. I had new water fountains and enough funds to install them so that there was no burden on the school. Unfortunately, we were met with some major problems. The school district first went through many details about which water fountains they could have so we made sure to get the right ones. But the bigger issue was that no regular plumber could do the work; they had to be union employees within the district. So, to upgrade the water fountains, I was told we would need to pay \$120,000 (versus the about \$3,000 we had budgeted). I tried absolutely *everything* to bypass this, but I had no luck.

Even the *Los Angeles Times* offered to publish an article written by a student named Andrea Hernandez, who was a bright young girl in our program. It would be published in their "High School Insider" publication, which geared toward a community of young readers, thinkers, and storytellers and creates opportunities for young people to develop their writing, media literacy, and multimedia skills, connect with *Los Angeles Times* staff, and immerse themselves in the field of journalism. Below is what she wrote.

Featured

Los Angeles High School

The impact #LAGREENTEEN has had on students

Andrea Hernandez

June 18, 2019

When Habits Of Waste first came to Los Angeles High School to educate my classmates on the negative impact that plastic has on the environment and to inform us about the #LAGREENTEEN movement, numerous kids were a bit hesitant. My classmates and

I thought "How can we reduce the use of plastic, when plastic is a part of everyday life?"

After guidance from HOW my classmates and I realized that there are various alternatives to plastic products and that there is in fact a way to reduce plastic use. However, as of right now students at L.A. High have no other option but to bring plastic water bottles from home or purchase plastic water bottles from the student store being that the water fountains at our school are unsanitary and outdated making the water from the fountains "undrinkable." #LAGREENTEEN is helping fix this issue by raising money to give L.A. High School new water fountains with a water filtration system.

The hope is that the new water fountains will essentially eliminate the use of plastic water bottles at L.A. High School and will also give students cleaner water to drink. The #LAGREENTEEN movement teaches kids about environmental care and how to bring awareness to others through the power of social media. This movement has given my classmates hope that we can save our environment for the future generations to come. Despite the fact that we're a relatively small group my classmates and I are very confident that we can bring awareness on the impact plastic has on the environment through social media.

My classmates and I are extremely excited about the work we are doing with HOW and #LAGREENTEEN as well as the new changes that are to arrive at our school. The majority of my classmates are already bringing awareness to their friends and followers on their social media's. Which has led to many students replacing their plastic water bottles for reusable bottles.

Both HOW and the #LAGREENTEEN movement have opened up my eyes and made me appreciate the world that I live in, much more than I did before. Before learning in depth about the effects plastic has on the world, I continuously used plastic products. Now that I'm informed, I've taken measures to reduce my use of plastic as well as my family's use of plastic. Being able to participate in #LAGREENTEEN has also made me think about my future as well.

I wasn't too sure what I wanted to do in life, but my experience with #LAGREENTEEN has made me want to pursue a career that'll allow me to make a change in the world. HOW and #LAGREENTEEN have given me a sense of power and direction. I've learned that anyone, regardless of age or race can make a difference and impact the world.

> I now feel that if I set my mind on a specific issue and I'm truly passionate about it, I can do something about the issue and bring awareness to others. Despite the fact that I'm a shy person, I've made an effort to inform others about the effects plastic has on the environment. I'm more than happy to bring awareness to others and guide them to reduce their use of plastic. Being able to bring awareness on the effects plastic has on the environment has made me feel like I'm really making a change in the world.
>
> To be a part of that change is an amazing feeling that I can't even begin to describe, all I can say is that I'm proud of my classmates and I because I know that we're making an impact on the world that will greatly impact, our generation as well as the future generations to come.

Reading this young girl's article forced us to keep going despite all the red tape we faced from the district. While I could not meet the $120,000 demand from the district, I decided that I would keep making noise about this issue. This is sometimes the only way forward.

We took this situation to the State of California to push for legislation demanding that all schools have access to clean water. California Assembly Bill (AB) 2638 was introduced by Assembly Member Richard Bloom (CA-50), whose field representative had taught the students alongside us in the #LAGREENTEEN program. Clearly we had struck a nerve within his office as well. (Coauthors include Marc Levine [CA-10] and Robert Rivas [CA-30]). The bill's goal was to provide water bottle filling stations in all K–12 public schools in the California. I was among the many who spoke at the assembly hearing, and I felt so proud to once again share the video that BuzzFeed helped create as a tool to convince legislators to make the much-needed change. Many environmental nonprofits like mine, as well as many doctors, needed to step in and explain the obvious health benefits of clean water for our students. The real truth is that having clean water is imperative for student success; it not only encourages hydration and provides a healthy alternative to sugary sports drinks but also supports memory and attention, as well as overall health.[6] We are grateful to everyone involved. The bill also states that K–12 public schools "shall *encourage water consumption through promotional and educational activities and signage that focus on the benefits of drinking water and highlight water bottle filling stations throughout schools.*"

This is important because studies show that children are most likely to consume more water in place of sugary drinks when drinking water is not only made more accessible but also actively promoted.[7] I think we owe this part to the many doctors who were pushing for AB 2638 to pass. Governor Newsom signed into law on September 29, 2022, after it received unanimous approval in the California Assembly and Senate. We were proud to be a part of AB 2638, and now I can sleep at night knowing that the students at Los Angeles High School will have access to clean water no matter what.[8]

In March 2023 LAUSD announced the Drinking Water Quality Program and published a statement to announce the approval of $33 million to "continue efforts to reduce and maintain the level of lead in all school drinking water to below 5 parts per billion—one of the strictest requirements in the nation for a school district." This allows for "the remediation of drinking water fountains and [installation of] water bottle filling stations at all special education centers and elementary schools. As part of the current phase of the Drinking Water Quality Program, the District has completed a comprehensive resampling of drinking water fountains at all K-12 schools, and to date has completed remediation work at 183 sites serving the youngest students and most sensitive receptors, including all early education centers."[9]

#RETHINKTAP

I wanted to understand the root cause of this issue and our dependency on single-use plastic water bottles. How did it become the norm to sell water out of a plastic water bottle? Where did this cultural phenomenon begin? I really wanted to understand the cause so I could help find a solution. It was crystal clear that there was and still is a deep mistrust in our tap water. Here in the United States, it should be assumed that all our water is clean and safe to drink. Sadly, as we all know, there are many cities where this is not the case (e.g., Flint, Michigan). The images of brown water coming out of the tap are just too overwhelming, and rightfully so. The truth is that while there are a few cities that do not have safe drinking water, there are *many* that do—almost 95 percent, in fact.[10] However, there is another reason why tap water hasn't been able to compete with the multi-billion-dollar bottled water industry. It's all about marketing.

I compare the beginnings of the plastic water bottle industry to the beginnings of infant formula. You see, the infant formula industry all began by sending very chic and beautiful women into groups of new moms to share that breastfeeding was passé and not as nutritionally sound as formula. They would convince well-off new moms to dry up their breast milk because they could afford the "superior" infant formula, and that lower-income people were the only ones to breastfeed because they couldn't afford the alternative.[11]

Today we all know what a farce that marketing ploy was. But guess what, it worked. In the 1960s and early 1970s breastfeeding took a nosedive, and many women would dry out their breast milk with medications such as Parlodel right there in the hospital since they had already started their newborn babies on infant formula.[12] Luckily in the early 1980s women retaliated with the "Breast Is Best" campaign and reeducated everyone (including some doctors) on the benefits of breast milk.[13] This campaign was another grassroots effort to make breast milk cool again. Whoever thought that something as natural as a mother's breast milk could go in and out of style with a strong marketing campaign? The reality is that we are facing a very similar effect today around plastic water bottles.

Cut to the 1980s during New York Fashion Week. Many glamorous and beautiful models walked fiercely down the catwalk for the most high-end fashion houses in the world. In this decade, models started carrying a very unusual accessory: a large plastic water bottle by Evian. Yes, a plastic water bottle was a fashion accessory.[14] No question about it, these marketers are smart. They used fashion models to launch a campaign promising that the water sold in these plastic bottles was superior to the filtered tap water we were all drinking. The marketing campaign promised to make you more beautiful and slimmer, have better skin, and be healthier—essentially, be like those models. Evian set the stage for many plastic water bottle companies to promise similar things, changing it up every so often with added electrolytes or something like that. Not to mention that A-list celebrities like Jennifer Aniston, Patrick Dempsey, and Pete Davidson, to name a few, who would become their spokespeople and join the countless other celebrities photographed with these bottles.

At the time many people balked at the idea of paying for water. It was seen as yet another ridiculous fad that would surely go out of style. "What's wrong with tap water?" they asked. Nothing. Absolutely nothing. While there are some cities in the United States that have terrible tap water, most cities are actually quite proud of their water. In fact, every city must provide a water report stating exactly what is in the tap water, and most times this is quite similar (even better!) than what you would find in bottled water.[15] I can imagine readers feeling alarmed right now, but tap water is actually much more stringently regulated than bottled water. The EPA regulates tap water, while the FDA (Food and Drug Administration) regulates bottled water (more on this later).

You can be sure that plastic water bottle companies are continually creating doubt in our minds about the safety of tap water, even though *many* of them are literally just filling their bottles with tap water, so their customers are simply paying for the plastic bottle.[16] Cities like New York developed strong marketing for their tap water, calling it "New York's Finest." As a New Yorker you are actually expected to drink the city's tap water because it's a source of pride for the city. Most cities don't have the budget or expertise to market their tap water, so you will see basic postcards with a blue water drop shape coming out of a kitchen faucet, as opposed to the flashy, multi-billion-dollar advertising campaigns that water bottle companies develop to convince us that their water is superior.

This is when I created the campaign called #ReThinkTap with the idea that many cities are proud of their tap water yet cannot find a way to communicate it effectively to their community. We created free marketing materials for cities to adopt and share with their residents. These materials are linked to the city's water report, which is imperative to look at because, as mentioned above, a city's water quality is regulated by the EPA rather than the FDA, which regulates the water bottle companies. The EPA is much more regulated than the FDA, thus making tap water safety requirements far more stringent.[17] We believe that this information should be more accessible to the public, as cities nationwide work so hard to ensure safe tap water and are required to make their water reports public. Our goal is to help cities entice people to at least learn what is in their tap water so that they can decide whether or not purchasing bottled water is actually necessary in their community. We want people to ask their city questions and take advantage of the water report.

We are all subject to the marketing campaigns of major companies, as they shift culture and society. Just think, they managed to get us to use 1.2 million plastic bottles per minute worldwide.[18] They started saying that tap water wasn't as good as plastic-bottled water, and many of us believed them. My goal is to have you ask questions and see for yourself. Why are we spending so much of our hard-earned money? (As of 2015, we were spending an average of almost $300 per year on plastic water bottles.[19]) Wouldn't it be better to do something else with that money, especially for those who face food insecurity? Imagine if they could use those funds to feed their families healthy nutritious food.

We can shift culture too, but it takes more effort, as millions of people must buy into the change. But we must start somewhere, and slowly the ripple effect will take hold and we will begin hearing about the changes happening worldwide. Just look at what the City of Paris has done with the hydration stations all over the city. These stations even offer sparkling water! The City of Los Angeles is installing and refurbishing two hundred refill stations just in time for the 2028 Summer Olympics.[20] We can begin a whole new infrastructure if legislators hear our cries and concerns. Don't lose hope—just take action, one step at a time.

HOW YOU CAN BE AN IMPERFECT ENVIRONMENTALIST

1. Take a look at your water report. If you have questions, call the number listed on the water report.
2. Take action and see what you can do in your home. Perhaps try filtering the tap water . Remember, many water bottle companies don't sell water; they sell the plastic that it comes in. They are not creating water; rather, they are getting it from natural resources, similar to our tap water.
3. Ask questions. If you see something that just does not make sense, ask questions. Reach out to your local representative's office and share your concerns. Sometimes they will take the lead, just like Representative Richard Bloom's office did.

4. Perhaps your school needs new hydration stations. Well, that can be resolved. Get the students involved first. Have them take photos, post them on social media, and then begin putting pressure on the school district to budget for the upgrades. It's imperative for our children's health, in addition to mitigating plastic waste.
5. Access to clean water is a human right. Gather in your community to ensure this is a priority for your city council. This should be a priority for every city council.
6. Speak to your local high school science department or administrators and bring the #GREENTEEN program to your students. At www.habitsofwaste.org we have all the materials they need to learn how to get started on making change.
7. Don't fall for marketing tactics. Do your research and learn what you can on your own before assuming the special water ingredients you hear about in ads are actually necessary for your health.
8. Download apps like WeTap Drinking Fountain Finder to easily learn where you can refill your water bottle with filtered water for free. Many local restaurants offer water for your bottle at no charge.
9. Make it a habit to always leave your home with a full reusable water bottle. It's a great habit to start forming so that you don't get stuck buying a plastic water bottle.
10. Reusable water bottles aren't just for water. Get creative and make your favorite iced tea or lemonade at home to take with you. You will save money while helping reduce plastic consumption.

5

#SHIPNAKED/#SHIPGREENER

Online shopping is here to stay. The convenience makes it possible for many of us to keep up with our everlasting to-do lists and allows for a speedy way to get things done. As you may know, shopping online comes with a huge carbon footprint, but there are ways we can do better. Shipping products (and returning them) produces 37 percent of greenhouse gas (GHG) emissions globally. Packaging also contributes to e-commerce GHG emissions, from deforestation for cardboard boxes to emissions from plastic packaging production.[1]

You see, the idea of putting one item in a box to be sent to someone's home is a very new phenomenon. Historically, when we shopped mostly at brick-and-mortar stores, a whole division of staff members was responsible for unpacking large shipments where thousands of items would arrive in very large containers and boxes. The staff members unpacked these boxes and placed the items on the shelves for sale. Afterward the boxes were compressed and properly placed into the recycling bin, and the whole system of recycling cardboard boxes was made possible. This is because the boxes were properly placed in the bin (i.e., without tape and other fillings that are not recyclable) and in one stop the sanitation team could potentially pick up hundreds of boxes to go to the recycling plant.

Suddenly we shifted our culture so that billions of boxes are arriving at individual homes while brick-and-mortar stores are being replaced with online shopping giants. Currently, Amazon sends out over 7.7 billion packages to homes each year.[2] Of course, their goal is to ensure the items get to our homes in one piece, so they overpack and repack the items in so much excess plastic packaging that it sometimes looks like an April Fool's joke. I am sure you have opened a large box containing one tiny, unbreakable item surrounded by plastic air pillows or air bubbles.

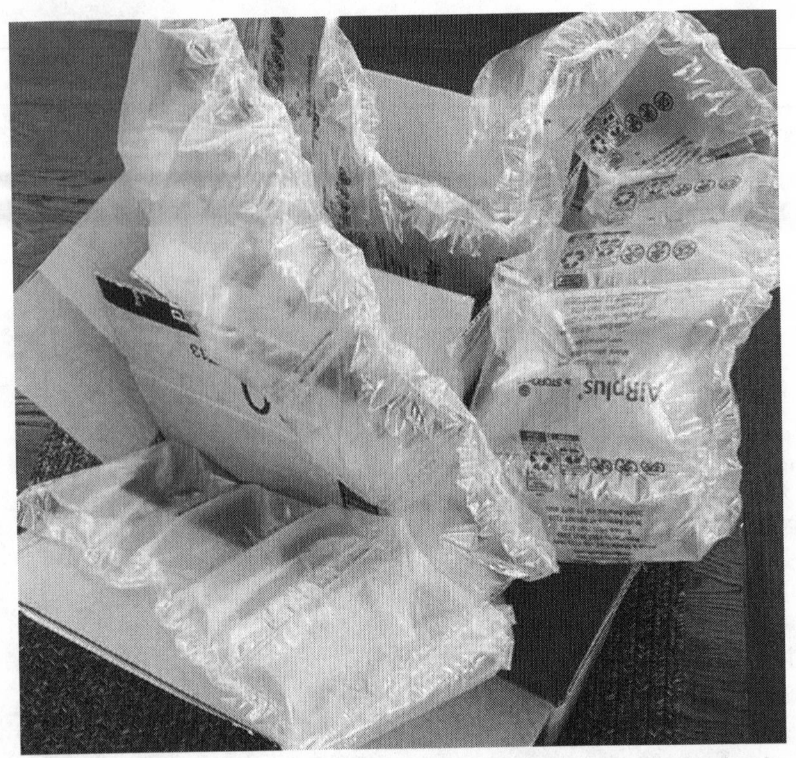

Figure 5.1. An Amazon shipment I received with excessive packaging for the size of the item ordered. Photograph by Sheila Morovati.

This always bothered me, so often I would avoid online shopping. But with two kids and working on my charities full-time, I was often left in a pinch and begrudgingly reverted to the ease of ordering from Amazon or the like. One day something strange happened. I saw a box sitting on my porch and it had a photo of a handheld vacuum on it. At first, I wondered if one of our friends had hand delivered this to my home as a practical joke because of the running joke that my husband is the cleanest guy around and loves gadgets like that vacuum. But as I took a closer look, I noticed that it had a shipping label on it and was sent to us from Walmart. It was my husband's new toy, and it was something he had seen online a few days earlier and ordered.

I asked if he had noticed upon checkout whether there was an option to ship the package "naked," but he said there was nothing to click on to receive the item without extra packaging. The great thing about this package was that the manufacturer had done an excellent job of securing the item, so placing it within another cardboard box would have been redundant and unnecessary. For some, the downfall would be security issues since you could definitely see what the item was, as it was pictured on the outside. But I was thrilled to know that this could be an option for people who might ship to Amazon lockers or live in an apartment building or a house where they didn't have security concerns. This is when I was inspired to create the campaign called #ShipNaked.

The truth is, online shopping is really harming our planet. Currently, we cut down the equivalent of one billion trees per year to provide enough cardboard for just our Amazon orders,[3] not to mention the plastic pillow packaging that Amazon uses so much of that it could wrap around the planet five hundred times.[4] It's just unsustainable to the maximum degree.

At the time, I imagined that shipping "naked" could also benefit the retailer, as there could be a cost savings. They would have to purchase

Figure 5.2. Behind-the-scenes footage of the campaign video for the #ShipNaked initiative. Image courtesy of Habits of Waste.

fewer boxes and less filling, automatically making the packages smaller, so the shipping would also be cheaper. I always try to put myself in the shoes of the people/corporations I am trying to nudge to see how I can convince them to care. While I care deeply for the planet and understand that perhaps some people working at Walmart, Amazon, and Target do too, it is clear that their bottom line is what they care most about, especially in public companies. In order for me to make any inroads with these companies, I know I need to sell them on my ideas using language they care about and means that provide them a savings in dollars and cents.

If I could prove that there is a cost savings, then I had another win-win campaign, which was a big part of #ShipNaked. I wanted to get the public's attention quickly, so we produced a PSA (public service announcement) with none other than a semi-naked deliveryman. How we pulled this one off is really one of those questions I often ask myself. If you put your ideas out into the world, many times the people around you can help you make them happen. My first call was to my executive director, Heather Lazarus, to share the idea about #ShipNaked and that I wanted to have a naked deliveryman drop off a toaster in the manufacturer's packaging. She laughed at first but then realized that I was serious. She then proceeded to tell me that her kids' soccer coach makes extra money by modeling and would likely help us out.

The next thing we needed was a director/cameraperson. Coincidentally, a man by the name of Jordan from Kindhumans.com was coming to my house the following week to capture content and interview me for their website. I had seen his work before, so I asked if he could stay an extra hour to create this thirty-second PSA for us. He also thought it was really funny and enthusiastically agreed. In less than a week we called, "Action!" and had Pasha, our new friend/soccer coach/model, in a Speedo carrying a toaster up the stairs of my porch.

On the day of filming, as luck would have it, I received one of the most atrocious plastic-filled, extremely unsustainably packed boxes from Amazon. I had Jordan film me opening the package and pulling out what seemed like an endless stream of plastic pillow filling. It all came together so nicely, and within ninety minutes we had filmed another one of the most provocative environmental PSAs in history. With this PSA we gained coverage from *Vogue Business*, BuzzFeed, *Business Insider*,

Figure 5.3. Feature on Amazon where the customer can opt to receive their order with reduced packaging. Image courtesy of Habits of Waste.

and ATTN: media, among many other outlets. We started to make the noise needed to get the attention of the online shopping giants.

In conjunction with the release of the PSA, we created another online email campaign targeting Amazon, Walmart, and Target and asking that customers be given the option at checkout to have their items shipped "naked" when possible. The email we used did not shame or blame them but expressed that their customers were tired of all the excess packaging, and that it was time to think differently. It was true though. Everyone I knew shared the same frustration and guilt when shopping online and would come to me with the hope that I would magically be able to make this "habit of waste" go away. It was time to give customers an option to do better. Although the name of the campaign was very racy, it very easily communicated the common concern that many packages don't need another layer on top of the manufacturer's packaging.

Sometimes taking one step forward is all you need to build more momentum. We thought it would be a good idea to reach out to UPS, the largest package delivery company in the world, to see what they thought of our #ShipNaked campaign.[5] I searched my LinkedIn profile and found an old friend of mine who still worked at UPS after more than twenty years. She loved the idea of #ShipNaked but immediately explained that there would be no chance that a huge corporation like UPS would support a campaign with that name, which is understandable because the word *naked* is triggering for some. We were disappointed but not for long because, as they say, when one door closes another one opens.

I realized there was an opportunity to actually expand the campaign we had built since many items we buy online don't come in rigid packaging from the manufacturer. Some items we buy may not need any excess packaging or even a huge box, yet they continue to arrive swimming in a massive cardboard box that is overflowing with unneeded plastic. In fact, I had just ordered a memory chip for my son's underwater camera, and it came in a 24-inch by 16-inch by 4-inch box with 3 feet of plastic pillow stuffed in there. The microchip was already packaged tightly and safely in a plastic envelope and did not need all this extra packaging. Perhaps it could have been placed in a small envelope or something of the like.

What a great chance to address all these other unsustainably packed items too. With that #ShipGreener was born. (I used hashtags in the names of these campaigns so that people would know there was a simultaneous social media movement that we hoped would go viral.)

#ShipGreener is meant to help reduce packaging for all shipments that cannot be shipped "naked," or in the manufacturer's rigid packaging. The name of this campaign also helped us work with more companies like UPS. I was happy to know that UPS had a whole team of people to help retailers ship their goods in more economical and eco-friendly ways. This group is called UPS Customer Solutions, and I recommend retailers take a moment to see the support UPS offers.

UPS embraced the #ShipGreener campaign and interviewed me for one of the case studies published on their website. #ShipGreener also affects their bottom line. Here's why: more compactly packed boxes allow for more boxes to be loaded in their trucks for delivery, which means fewer trucks on the road, and that equates to major savings in fuel costs and fewer drivers needed. This ripple effect begins with campaigns like #ShipGreener, so we had to keep pushing. Once again, speaking the language of these corporations was my ticket to driving change. The bottom line (financially speaking) is something they care deeply about, in addition to their ESG goals. So, if we want these huge companies to take action, we must find ways for them to benefit financially, while providing them with a sound solution for problems.

Building a social movement is a dance. We need to put pressure on the top, but at the same time, we need to empower individuals to express the need for change. After all, we are the customers of these mammoth companies, and without us, there would be no companies at all. To create the pressure from the bottom up we not only asked people to send the prewritten emails targeting the online shipping giants, but we also launched a social media campaign called #ShowUsYourPackage to ask individuals to take photos of online orders that come with excessive packaging. The next step is for the individual to tag the online retailer and request that they make a change and #ShipGreener next time. Corporations want to avoid being tagged online for anything that is negative, so it's one of the fastest and most effective ways to get their attention. Social media is our best friend, and we can really make our voices heard if we expose companies in this way. Instagram alone has 1.39 billion monthly users and TikTok is now at one billion monthly users, so imagine if everyone used their social media voice once or twice a week for something environmentally focused.[6]

We have seen significant change since the beginning of the campaign. For example, Amazon now has a toggle that offers users the choice to receive their item with zero or less packaging.

In addition, Oceana worked very hard to pass a bill in California to eliminate plastic packaging, but ultimately it did not pass. This bill, AB 2026, sought to restrict e-commerce shippers' use of single-use plastics, placing limitations on use volumes and subjecting violators to penalties.[7] I learned through that legislative effort that Amazon had already eliminated or reduced plastic packaging in many other countries, such as India and Germany, but had not even attempted it here in the United States.[8]

So we know that Amazon does better in other countries but chooses to continue with plastic here in the United States.

My hope is that Oceana will try again. They need the support of the masses to make this happen though. Legislators will take swift action when they know their constituents are fed up and want change. As with the BuzzFeed videos for #CutOutCutlery and #LAGREENTEEN, we were able to convince Sacramento that millions of people cared about this issue. Although we offered repeatedly, our naked deliveryman PSA did not make it to the assembly floor due to its racy content. Large organizations probably did not want to share that imagery. Luckily, Amazon is listening. We have sent thousands of emails to their corporate headquarters through our P2A platform.

Convenience is king, and people will continue looking for easy ways to save time without intentionally trying to generate more waste and harm the planet; they are just trying to get their needs met quickly and efficiently. We must demand that the largest companies lead by example and ship greener, especially if they are capable of doing so in other countries. We must hold them accountable by expressing our concerns in every way possible.

We want environmentalism to become a topic that everyone can get behind without feeling as though they must identify as a bona fide "environmentalist." These labels hinder people from feeling that they have permission to participate in the environmental movements needed to combat climate change. My belief is that if you live on this planet, you are automatically an environmentalist (no label needed, though!). We all depend on the health of our planet, as it is our only home. We must care for it the same way we care for the homes we live in.

HOW YOU CAN BE AN IMPERFECT ENVIRONMENTALIST

1. First and foremost, try to buy secondhand items as often as possible. These items are so much fun to find. The local flea market has so many treasures. And don't forget about the wide array of secondhand stores popping up everywhere. Enjoy the hunt!

2. When buying secondhand isn't a viable solution, try to physically go to the store to pick up what you need. This way, the items don't have to be repacked with plastic and cardboard and then shipped on trucks to reach you.
3. When you do order online, select the option to ship more items in fewer packages. You may receive your order one or two days later, but it will leave a lower carbon footprint and require much less packaging.
4. Participate in our #ShipGreener/#ShipNaked P2A campaign. Send your emails out so that we can continue to put pressure on the largest online retailers in the world. Visit www.habitsofwaste.org.
5. When you do receive an unsustainably packed item (e.g., a huge box with a small item inside surrounded by an excessive amount of plastic fillers), take a photo, post it on social media, tag the company you ordered from, and use the hashtag #ShowUsYourPackage so that we can see it and repost.
6. Be vocal. If your item arrives unsustainably packed, contact the customer service department of the company you ordered it from and explain your concerns. They are all listening.
7. Save the cardboard boxes! Post on nextdoor.com to inform your neighbors that you have many boxes in case anyone is moving. The likelihood of those cardboard boxes getting recycled is slim, so let's help our neighbors out instead.
8. If you do place your boxes in the recycling bin, keep in mind you must remove the tape in order for it to be properly recycled.
9. Find out if your city will recycle plastic fillers and plastic envelopes. Chances are they won't, but you can spend a few extra dollars to get them recycled by private companies like TerraCycle.
10. Shop local. Try to purchase items that are manufactured nearby. It makes a big difference.

6

#BARSOVERBOTTLES

PHASE 1

Let's take a moment to envision what virtually every shower in America looks like. You will likely find a corner nook with liquid shampoo, conditioner, and body wash, all in plastic bottles. These are just some of the essentials for our personal hygiene that we all regularly purchase. It's such a massive part of life that the beauty industry rakes in over $100 billion in revenue each year with just under $8 billion spent on advertising products.[1] This massive industry is the cause of so much pollution.

Plastic has become such a major part of our lives that much of it goes unnoticed. For years, I guiltily looked around my shower because I knew I shouldn't have all those plastic bottles. There were two for my hair (shampoo and conditioner), one for my body wash, another for my face wash, and then those for specialty products like exfoliants and deep conditioners. I just turned a blind eye for decades because I couldn't find another option, nor was I at a point in my life to make my own shampoo (I still haven't reached that point).

One day I heard this statistic: In the United States alone, 550 million empty shampoo bottles are thrown out per year, and the average American goes through twelve bottles of shampoo a year.[2] As a reminder, less than 9 percent of plastic waste actually gets recycled.[3] When I hear numbers like this, it really alarms me and forces me into action. Generally, my rule is to start small so that I don't get overwhelmed. I tell myself to take just one step at a time but continually move forward and make progress. I began devising my next plan to combat plastic waste. I wanted to once again learn the root cause and see if I could change my habits easily so that I could convince others to join in. I realized that up until recently no one had bottles of "body wash" because we only had bars of soap. So, as my first move personally, I switched back to bars of soap. Next I was reminded of my teenage years, when I bought my first face wash set from Clinique, which was always a bar of soap. So I decided to make that swap too.

Later that day, I stopped by my local pharmacy and local beauty supply store to make my swaps but left empty-handed, as there was virtually nothing in bar form. Instead I found aisles and aisles filled with a plethora of colorful options that were all packaged in plastic. Yes, there were still a few bars of soap available, but far fewer than the number of plastic options. Plus, most of the bars of soap that were available seemed like industrial-grade products.

I continued my search online and found many great options for wonderful face washes, body soaps with wonderful ingredients and scents, and a wide array of shampoo bars. The shampoo bar was a new concept for me, one that I was very curious about, so I went ahead and ordered the brand that had the best reviews.

Turns out the products I chose were excellent and I was thrilled. I could not tell the difference after washing my hair and face with the bars, so I knew this was a switch I could commit to. There was one issue, though. I did have to do more work hunting these items down, which was a red flag for me. Anytime I try to ask the general public to change their behavior, I want to be realistic and see if others who are racing around in their busy lives could also take these extra steps.

The reality is that we cannot expect people who are trying to just survive life to make things more complicated. Going online to search high and low for more sustainable options that at times still cost more than the plastic-bottled versions is not a realistic expectation we can have.

This is where we need corporations to step in and do better. As I became more and more aware of the beauty aisles in major retail stores, I realized there are just rows and rows of plastic that we will buy and eventually discard without even having the chance to take a more sustainable alternative path.

It was time for more bar alternatives or even the sustainably packaged options to have a place on those shelves. I created new campaign called #BarsOverBottles, which came with a P2A email campaign geared toward asking the four main shampoo manufacturers to spend

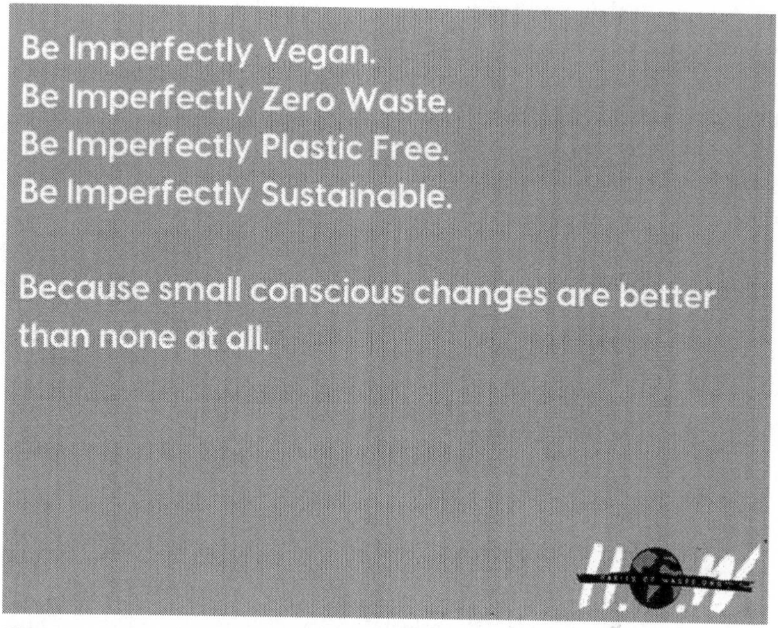

Be Imperfectly Vegan.
Be Imperfectly Zero Waste.
Be Imperfectly Plastic Free.
Be Imperfectly Sustainable.

Because small conscious changes are better than none at all.

Figure 6.1. Be an imperfect environmentalist! Image courtesy of Habits of Waste.

some of their research and development budget on creating a more sustainable shampoo in bar form.

We didn't want to reinvent the wheel, so here is an example of our request. America's favorite shampoo brand is Pantene, which was created by the behemoth Procter & Gamble. Our ask was for them to spend some of their resources in research and development to create Pantene shampoo in bar form or more sustainable packaging. If we could show Procter & Gamble that the public would purchase their products in a bar form or more sustainable versions, then perhaps they would be willing to join in and make the change. Yes, there are many layers to consider, as a bar form of Pantene would likely change the look of the product on the shelf, but there is no time to waste. We need to reimagine the industry together so that the planet is made a priority over what looks better on the shelf in order to increase sales. By providing customers with a more sustainable option, we can at least begin to move in the right direction and reduce our deep reliance on public enemy number one: the fossil fuel industry, where all this plastic is derived from in the first place.[4]

If sustainably packaged options were readily available, then we would have a stronger case when trying to push individuals to change their habits. We would reiterate over and over that individuals can support sustainability with their dollars and show these companies that we as a culture are ready for these new options. Companies study our behavior patterns by analyzing what we purchase, so my goal is to get the public to support the new eco versions of our favorite beauty products if and when they are made available to us. It would be our way of saying thank you to the corporations' efforts to be greener. The bottom line is that we must provide society with options in order to suggest alternative behavior.

We are still in the first phase, which begins with an email campaign for #BarsOverBottles giving consumers the chance to reach out to the biggest companies in the beauty industry. Our email is currently directed at the sustainability and social impact departments at L'Oréal, Procter & Gamble, Unilever, and Johnson & Johnson.

Below is an example of the #BarsOverBottles email that we drafted to Unilever. This is written by us for individuals to send by simply entering their email. This is beneficial because the corporations we are trying to reach will receive thousands of emails from individuals using their unique email accounts. The impact is very powerful, especially when you receive the same email over a thousand times from a thousand different people each month.

> Dear Unilever Director of Social Impact,
>
> Our goal here at Habits of Waste (HOW) is to remove obstacles so that individuals can more easily make better choices to help protect the environment. As part of our #BarsOverBottles campaign, we are requesting that your incredible development team create a bar version of your shampoos and body washes. We realize that your products are loved by Americans, which is why we believe that working together to create better products is more effective than trying to boycott items with plastic packaging.
>
> Every year, over 120 billion pieces of plastic packaging are produced by the beauty industry alone. These products do not decompose; rather, they endanger wildlife and harm our oceans and human health when disposed of.

We know there is a future in package-free products, as it is a fast-growing alternative. We invite you to be a leader and a pioneer by moving toward more environmentally sound products that would reduce costs related to production and shipping due to their lighter weight. Your commitment to the planet will allow your customers to continue purchasing their favorite products and reduce their environmental impact.

Please join us and invest in creating a plastic-free, more sustainable bar alternative to traditional personal care products.

Thank you for your time and consideration. This switch has the potential to completely disrupt the personal care industry and make Unilever a leader in sustainability. We encourage Unilever to join in and help lead the way for other companies and corporations—the environment needs you to pave the way to heal the planet. Please visit @howchangers and www.habitsofwaste.org for more information.

Sincerely,
[Your Name]

We are still working toward our ultimate goal to significantly reduce plastic packaging in the personal care industry. The #BarsOverBottles campaign is our most challenging yet because we still cannot prove the bottom-line profits that corporations will gain with this change. Without the alternatives available to the masses, we can't attempt to show that people will purchase the bar form. Eventually, with enough people participating, we will begin to see the shift come to life.

I often talk to my team about looking at an eight-burner stove with pots going on each of these burners. Each of these pots represents one of our campaign ideas. As a pot gets stronger, it is moved to the front of the stove. #BarsOverBottles is a campaign that is gaining momentum. Recently, I was able to connect with L'Oreal's sustainability department and was happy to see the progress they are making toward more sustainable packaging, refillable options, and even more bar forms of shampoo. L'Oreal's global director of sustainable packaging and development, Brice André, made the following statement: "Our efforts to reduce packaging intensity, replace materials with circular ones, conceive packaging so that they can be reused, . . . develop refillable packaging, and design packaging so that they can be recycled. We do not forget breakthrough steps that enable us to reinvent packaging such as solid shampoos."

We believe that in time and with a little bit more persuasion, we can influence the beauty industry to try new ways of packaging their products. There is even an EcoBeautyScore Consortium that includes seventy major beauty companies. Their goal is to agree on new methodologies for sustainability, which is a breakthrough industry initiative. We are on the right path, but we wish things would move along more efficiently, as there is no time to waste.

One thing we are seeing that leads us to believe that change is near is how much energy and effort is going into alternative laundry detergent packaging. We are seeing pods of detergent, sheets of detergent, and more inventive solutions to steer us away from all the rigid plastic containers, which barely get recycled. Most of these end up in our landfills and our oceans. The truth is that the only reason for these large plastic containers of detergent or shampoo or body wash is the space needed for water to be mixed in with the soap to make it liquid. Liquid detergent can contain up to 90 percent water, meaning you may be paying for just 10 percent detergent because the companies dilute concentrated detergent in order to fill up their big, colorful jugs to make it seem as though you are getting a lot of detergent. What you really are paying for is the expensive shipping due to the weight, while the packaging contains very minimal amounts of the actual soap that makes up the detergent.

New companies selling just the concentrated detergent without the water are popping up everywhere. The concentrated detergent is lighter and much less expensive to ship, so there is a major cost savings here. This movement is setting the trend for more change to come. It's up to us to support these environmentally focused companies and give them a chance to survive. We must also play our role by using our dollars to show that we are ready for these eco-friendly alternatives. Please consider giving one of these products a try next time you are on the market for laundry detergent.

PHASE 2

The next phase of this campaign involves addressing all the toxins found in our beauty products. Many of the products we are allowed to purchase in the United States are forbidden in other parts of the world, especially Europe, because of their potential health and environmental harms.[5] In addition to a number of health risks, these toxins also contribute to pollution of our ecosystems when they reach the water during our personal showering and washing.[6] We added a layer to #BarsOverBottles requesting that beauty retailers allot a section for nontoxic products so that we as consumers know that there are options available to us that are better for our bodies and better for the planet.

We have all heard about sunscreen and the damage it causes to our coral reefs due to the nanoparticles in them. We are glad to see more marketing efforts done to promote "reef-safe" sunscreen and that companies are proudly placing a sticker or notification right on the packaging. If you don't see the words "reef-safe" indicated on the front of the product, look at the ingredient list and avoid products with the two Os: oxybenzone and octinoxate. (Extra credit: Also look out for benzophenone-1, benzophenone-8, OD-PABA, 4-methylbenzylidene camphor, 3-benzylidene camphor, and octocrylene.) These can affect the coral reef's reproductive cycle, damage DNA, and worsen the effects of coral bleaching.[7]

I decided to create another layer to this campaign because I have personally met some of the most dedicated small business owners who work tirelessly to create toxin-free alternatives so that we can have choices; yet, simply because of money, the stores do not make space for them. It should be mandatory that we have access to these wonderful alternatives. Perhaps there should be a law stating that a certain percentage of beauty products sold from any store must be free of toxins. Otherwise, just like the shampoo bars, we have no opportunity to do better.

We must create the pressure needed for the United States to do away with all the chemicals they've been permitting. We want the United States to follow the path of banning chemicals from our beauty products, just like the Europeans have. (Lead, parabens, formaldehyde, and phthalates are just a few of the more familiar toxins that are banned in Europe but still found regularly in the United States.)[8]

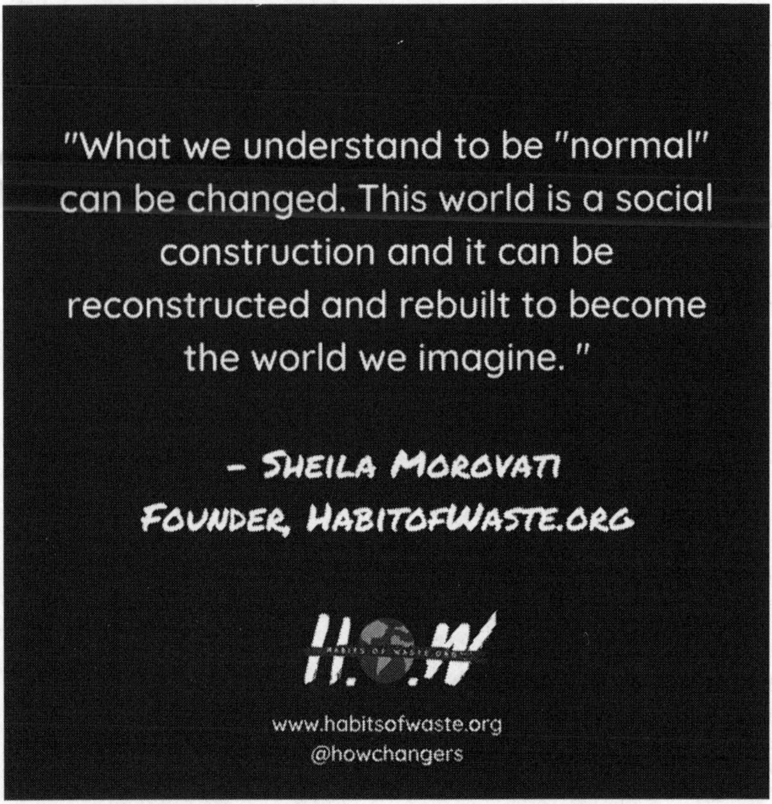

"What we understand to be "normal" can be changed. This world is a social construction and it can be reconstructed and rebuilt to become the world we imagine."

- Sheila Morovati
Founder, HabitofWaste.org

H.O.W
www.habitsofwaste.org
@howchangers

Figure 6.2. Challenge norms. Image courtesy of Habits of Waste.

I'll share one of my favorite stories about making the switch to a nontoxic product and never looking back. I learned about two women, Leah Yari and Mary Lennon, who after seeing a little girl's spa-style birthday party decided to create a line of nail polish that was the cleanest on the market, meaning it would be free of at least eleven of the main toxins. They called it Cote and started to make incredible progress by innovatively creating a product that had never been possible before. With a lot of hard work, they created their line of products without all the chemicals you normally find in nail polish.

I wanted to support their company, so I purchased a couple bottles of their polish and applied it to my nails. I wasn't expecting much because I naively believed that those ingredients they managed to eliminate were the ones that made nail polish actually stay on nails. Much to my surprise, the polish didn't budge for an entire week. I was shocked to find a superior product without the toxins that left me with a whole week of beautiful nails without a chip or a crack. I have yet to buy any nail polish other than Cote, but it was no coincidence that I stumbled across this product. I personally knew both the women who created it. Had it not been for that connection, I don't know if I would have found out about this wonderful product that is safer for my body and the planet. The reality is that more people could be benefiting from products like Cote if there was room created for them in Target, Ulta, and Sephora. It's up to us to push retailers to give us access to amazing products that are out there.

"We've been doing it for years" is not a strong enough reason to continue down the same path. We have new innovations that are solutions to big problems so we can create new habits. You may be pleasantly surprised by the results and the quality, as well as the potential cost savings.

HOW YOU CAN BE AN IMPERFECT ENVIRONMENTALIST

1. Try a bar form of shampoo, soap, face wash, or body wash and see how you like it. You might be pleasantly surprised.
2. Join the #BarsOverBottles P2A campaign and send your email to the companies we're targeting. Visit www.habitsofwaste.org
3. Try switching to a concentrated form of detergent for your home. Those big jugs of detergent are a thing of the past, as many new options are becoming available to us.
4. Take photos of your new finds and share them on social media. You may influence a few (or many!) people to make the switch too.
5. Let your local store manager know that you would like them to carry more sustainable options for both beauty and home care items.

6. Help us communicate with major retailers that we want access to products that are free of toxins. We shouldn't have to hunt all over the internet to find good, clean beauty products.
7. Contact the Food and Drug Administration to implement stricter standards for beauty products in the United States (see fda.gov). You can even tag them with your concerns on Twitter (@US_FDA).
8. Use a reef-safe sunscreen!
9. Remember, what you put on your body eventually ends up in our environment, so buy clean products whenever you can.
10. There are many DIY (do-it-yourself) beauty products you can create at home that do not have any packaging whatsoever. Google "DIY beauty" and you will be amazed at how easily you can create a beauty ritual at home that is package free and low cost. Did you know that lip scrub can be made by mixing just olive oil and brown sugar?

7

#LIGHTSCAMERAPLASTIC

After the success of #CutOutCutlery, I was regularly included in meetings with different coalition groups trying to pass more legislation around eliminating single-use plastic. We were now one of the leaders in the space and felt like we could move the needle a lot, especially after receiving an award by the California Resource Recovery Association (CRRA) for our work. This is a very big honor because each year the CRRA awards a business, government agency, or community-based organization that has excelled in preventing waste or has implemented procurement practices that prevent or reduce waste. The CRRA is California's largest recycling association, and they lead the charge for waste management and reduction. Needless to say, all eyes were on us after the announcement of our award was made.

We continued to be invited to many meetings for legislative measures, and I started to get really bogged down by the massive scale of the problem, which was compounded by the toxic relationship between the plastic industry and the environmental organizations that were trying to ban plastic in cities and states nationwide. Here is just one example of what would happen when any progress was made: After ten years of tireless work, a single-use plastic bag ban was issued in the City of Los Angeles. Americans use more than one hundred billion plastic bags each year, which equates to more than three hundred bags per person per year.[1] Here are the details of this ban: "In 2013, Los Angeles became the largest city in the United States to ban plastic checkout bags. The reusable bag ordinance includes a ten-cent fee on bags as an incentive to remind the public to bring their reusable bags. The measure was a huge success and passed 11–1, proving the commitment of Los Angeles city council members with their action against plastic pollution."[2] Many of my fellow anti-plastic coalition members supported this ban, and major celebrations took place when it finally took effect in 2014.

Just when we thought we were done with plastic bags and on our way toward a cleaner future, the plastic industry and their savvy lawyers decided to pick apart the ban and find a major loophole. They took the words "single use" and strategized around that. They determined that they could produce thicker plastic bags that could be used more than once so that they didn't fit under the category of "single-use plastic bags" anymore.[3] California Attorney General Rob Bonta had concerns about this. Yet again, the plastic industry found a loophole. Most grocery stores turned to these thicker, "reusable" plastic bags, which they claimed were "recyclable." Thankfully, Attorney General Bonta went on to determine if they truly were recyclable as required by law in 2022.[4] He targeted the Novolex, Revolution, Inteplast, Advance Polybag, Metro Polybag, and Papier-Mettler companies and asked them to prove that their bags can be recycled in California. Last we heard it is still an "active and ongoing investigation."

Their sneaky strategy worked. To this day, when I enter the grocery store, I see hundreds of extra-thick plastic bags being given out. These newer, heavier plastic bags pollute our planet even more, but there is one silver lining. A new behavior throughout society was born. More people were adopting the new habit of bringing their reusable bags while shopping. We must continue to keep trying to augment the five hundred billion plastic bags that people use per year worldwide.[5] Needless to say, it is vital that we all take the necessary and easy step of leaving a stack of reusable bags in your car at all times so that they are always available to you.

The battle between the $5 trillion fossil fuel industry (i.e., the plastic industry) and their tricky lawyers against the anti-plastic coalitions is a cycle we needed to break. The link between fossil fuels and plastics is not clear to most people. You see, "plastics are part of a sector called 'petrochemicals,' or products made from fossil fuels like oil, coal, and gas. That's right, corporations make plastic using dirty fossil fuels."[6]

As I was watching television one evening I realized there was one critical yet achievable way we could win this battle, and that's when the #LightsCameraPlastic? campaign was born.

Movie night in my household was always a favorite, and it became more so during the COVID-19 pandemic. As for many people, it was one of the best ways to break up the monotony of being home for so many months. I looked high and low for great family-friendly films that my whole family would enjoy. Movie night was one of our happiest moments when the four of us would forget the chaos happening around us, pile on our big sectional sofa, and put on a film while munching on a variety of popcorn and snacks.

As we consumed more films each week than ever before, I started noticing so much single-use plastic in almost every film or television show I watched. An internal alarm began whispering inside my head. This was a major oversight by the entertainment industry happening right before my eyes. That alarm soon became a blaring siren forcing me into action yet again.

A teen-heavy show called *Cobra Kai*, which is a *Karate Kid* spin-off, includes kids drinking from plastic water bottles in almost every workout scene. In reality, most teenagers in high school have a reusable water bottle adorned with stickers personalizing their bottle as an eco-statement; they rarely drink from plastic water bottles. So was it for the advertising dollars? That could not have been it because the water bottles had no labels on them at all. Perhaps they were sending a message to advertisers to come and replace the plastic water bottles with one of their branded ones. In any case, a myriad of other water bottle vessels would have been more appropriate for depicting the current reality of teens while simultaneously denormalizing single-use plastic. A simple swap was all that was needed.

Another show we were watching was a Netflix series called *A Man with a Plan*, which also proves my point that Hollywood was promoting plastic way more than necessary. The show has a lunch scene where two ladies are eating lunch in a cafeteria. Each one of the ladies has salads in plastic containers, plastic forks, and sodas with plastic straws and lids. Behind them was an image saying, "Recycle here." The irony of that moment was almost too much for me to handle! Not only are they further normalizing plastic, but they are also implying that it can be recycled. Side note: It is estimated that less than 9 percent of plastics are recycled, and some studies suggest that in actuality less that 5 percent of plastics are being recycled![7] I took a photograph of this scene for our website.

Figure 7.1. Still from a Habits of Waste PSA from Mark Ruffalo in support of the #LightsCameraPlastic? campaign. Image courtesy of Habits of Waste.

I envisioned that same scene with reusables. Perhaps the characters could bring their salads from home in reusable containers and eat them using a reusable metal fork (salad tastes much better with metal forks anyway!). Or their cafeteria could offer a plate and a reusable fork for their salad. Their drinks could have been provided in a reusable cup or even a paper cup. The storyline wouldn't need to change in any way, shape, or form.

Then I started noticing the excessive amount of plastic even more. Oh, the amount of red Solo cups shown on-screen was unbelievable. I think it was serendipitous that we happened to watch teen films more frequently since I had a thirteen-year-old and an eleven-year-old at the time. The repeated visual of hundreds of red plastic Solo cups is what motivated me to start making a change. Now I understood why every college party I had gone to was exploding with red Solo cups. What we see on-screen determines what we believe is and is not acceptable throughout society. In the words of professor and scholar of psychology Dr. Franklin Fearing, "social scientists agree that there are profoundly important relationships between motion pictures and human behavior."[8] The images we see on television and in film impact our lives by solidifying cultural norms and values.

Something had to be done, and I had an innovative and fresh solution. Stats like "by 2050 there will be more plastic than fish in the ocean"[9] depress people and make us feel that we have already lost the battle and there is no hope left. I can only imagine what the public must feel when they hear that popular statistic. I'd prefer to go with the saying "out of sight, out of mind." Perhaps if we shift toward showcasing sustainable alternatives, our brains would begin to register those as normal and accepted. It's about retraining our minds throughout society and impacting the six billion people who watch films and television shows.[10] The chances to make change would be massive. I quickly remembered how not long ago we made smoking fall out of fashion by removing it from our screens thanks to the work of anti-smoking campaigns. This really inspired me because the results were profound. Research has proven that when characters smoked less on television, consequently adults in the United States also smoked markedly less.[11] I knew that if we used this as a model and removed single-use plastic from our screens, then we could achieve the same effect.

The first thing I wanted to understand was the process it took for a prop to make it on-screen. I decided to interview several prop masters and set decorators. These are the hard-working and very talented people who make the set come to life with all the details that create the scene the director is looking to capture. As I interviewed each one of these them, I asked over and over, "Why is there so much plastic on our screens?" Coincidentally, they all shared the same reasoning—they didn't feel they had the authority to create a "greener scene" with reusable props versus single-use options even if both could work equally. Their fear of getting into trouble with the director on shoot day was the main issue. They did not want the director to walk in and see a scene that was eco-friendly solely on decisions made by the set decorator or prop master.

This "fear" was causing filmmakers and the entire entertainment industry to unknowingly support the single-use plastic crisis. I explained to them that film and television subliminally tell viewers what values society generally believes.[12] By placing excessive amounts of single-use plastic in the hands of our on-screen heroes (i.e., celebrities), it is implied that plastic is cool or, at the very least, OK. I went on to share the studies showing how celebrities influence our decision-making processes, from what we wear to how we speak.[13] This is why so much money is spent on product placement in film and television. Companies pay to have their product included in film or TV productions in the hopes that the success of the production will beget success of the product.[14] This is a huge business—over $23 billion a year is spent on product placement. For example, picture E.T. following a trail of Hershey's Reese's Pieces in Steven Spielberg's *E.T. the Extra-Terrestrial.*[15] As a result, sales of Reese's Pieces candy increased dramatically. This represents the process of product placement, which happens to be a significant portion of the entire advertising industry, projected to be worth over a trillion dollars by 2026.[16]

It would also be fair to understand whether all this on-screen single-use plastic was ever really necessary for the storyline. It wasn't. There was absolutely no need for a dining scene in a nice restaurant to have a plastic water bottle front and center. What stood out to me immediately was that none of this plastic was relevant to the storyline. My first goal in launching #Lights-CameraPlastic? was to address the fear that the crew felt. I wanted to help neutralize the set so that anyone setting up a scene could automatically default to reusable alternatives without getting into trouble. I created a tool kit for any production company to download for free from our website. The first item in the tool kit was a prewritten letter for the showrunner, the executive director, or the executive producer to copy and paste into an email to all of their cast and crew during the preproduction phase (when everything is coming together before the actual shooting begins). Here is the letter I drafted:

> Dear Crew Members,
>
> For this production, we will be joining the #LightsCameraPlastic? campaign by the non-profit organization, Habitsofwaste.org. Our participation means that we will use reusable items on screen instead of single-use plastic items wherever possible.
>
> Why are we doing this? The numbers are staggering. Imagine 1 million plastic bottles discarded per minute and 500 million plastic straws discarded per day or 40 billion pieces of plastic cutlery discarded per year and 160,000 plastic bags are used per second. We

must find innovative ways to close the "tap." These single-use plastics harm our planet as they are not recyclable and take hundreds of years to decompose.

How can we help? Sociology and Psychology experts agree that film and television subliminally dictate what is and isn't acceptable in our collective society. Each film and television show created is potentially viewed by millions of people. What they see on screen normalizes human behavior. This is similar to the decline in smoking when it was reduced from Hollywood productions.

Together, we can create a ripple effect that inspires our viewers to be more sustainable in their everyday lives.

Thank you for your efforts to be a part of a big change.

Sincerely,
[Your Name]

This simple letter shows the problem of single-use plastic consumption and how their production, along with the entertainment industry as a whole, can take part in the solution. In addition, this email sets the tone and culture for the production since many people working on sets do not even know how bad the plastic pollution issue is, nor should we expect them to. The entertainment industry is just that—an industry focused on entertaining us through creativity and filmmaking. If organizations like mine can fold in environmental messaging into their art, then we can really work together to change the mindset of the world.

Next we created posters with QR codes that took people to an informational page explaining in detail why it is so important that we refrain from showing off plastics on-screen. This way, everyone on the production could understand the problem and why their effort matters and is actually imperative to change the culture we live in today. This ended up really working.

As per usual, I like to include the general public in our efforts, as well. We added an email campaign similar to #CutOutCutlery where people can send their emails to every Hollywood Guild and every major studio and ask them to commit to #LightsCameraPlastic? and stop showing single-use plastic on-screen for all their productions. This call to action gives non-Hollywood people the chance to make their voices heard by allowing them to email virtually everyone in Hollywood with one click. We target the Writers Guild, the Producers Guild, IATSE (International Alliance of Theatrical Stage Employees), the Screen Actors Guild, and the Art Directors Guild so that all the people who work in any and every part of the industry could hear about this effort and remove single-use plastic from view.

To further the effort, I met with several sustainability directors from a variety of studios and learned that they were actually totally on board with this campaign. For example, Sony Pictures had already disseminated a press release expressing their commitment to go totally plastic free by 2025, "building on [Sony's] progress and expanding our efforts to reduce the impact of content creation internationally while eliminating single use plastics."[17]

Their goal is to remove plastic on and off sets and everywhere else in their company. We are rooting for leaders like Sony to make this a reality. Our email campaign gave power to the general public to send in their request for studios to go plastic free in support of the internal efforts of corporations like Sony. We knew that this could help build the momentum needed to prove to studio heads that this commitment to refrain from showing single-use plastic is necessary and what people want.

Then I met several directors and producers who also believed in #LightsCameraPlastic? but felt very alone in an industry that had many other priorities. They appreciated the structure of the campaign and having the support of our environmental nonprofit pave the way for them to implement their eco-beliefs on their sets. *The Hollywood Reporter* found out about this campaign and wrote an article about it on July 15, 2021. We were thrilled and posted about it on social media. Within a few hours someone tagged a woman named Kat Coiro on Instagram. She was so happy to learn the campaign existed. She is a director and an executive producer who is on the front line of protecting our planet. We immediately had a phone call and she shared her similar passion to prevent plastic from being seen on-screen. In fact, her upcoming film *Marry Me*, starring Jennifer Lopez and Owen Wilson, was plastic free. She was happy to join forces and help us create a new norm throughout the entertainment industry. Kat helped us catapult the campaign forward. My team and I were thrilled. So it began.

BuzzFeed and *The Hollywood Reporter* continued to cover the growth of the campaign, which propelled the movement forward faster than we had ever imagined. Thanks to Kat, who is also the executive producer and director of *She-Hulk*, we were able to get the one and only Mark Ruffalo to be the opener for the BuzzFeed video and a major advocate of the campaign. The star power of Mark was another essential part of our success, and we are forever grateful. He even attended our advisory meeting for all sustainability directors of major studios to join a panel of experts on behavior change from the UCLA Department of Sociology and Georgetown University. After that event, Mark shared this quote with us, which we feel perfectly captures what we are trying to do: "Together, we can pretty easily move away from being $100 million commercials for single-use plastics." After all, why should Hollywood promote the harmful plastics/fossil fuel industry?

Mark felt we should create a petition for the "above-the-line" folks in Hollywood to sign. ("Above-the-line [ATL] positions refer to those responsible for the creative development, production, and direction of a film or TV show. Before sets can be built or cameras can begin rolling, these folks are responsible for guiding a project from idea to script to screen."[18]) Since the inception of our petition we have received signatures from Billie Eilish, Finneas, Christian Slater, Ethan Hawke, Fran Drescher, and of course Mark Ruffalo, plus many more. We are continuing to grow this list so that we can use these names as a strong force to show studios that plastic is out and reusables are in. Over time I eventually met with every single guild we had been emailing through our P2A platform and learned that the set decorators and art directors often want to make change as well but don't have the tools to convince the decision makers such as the director or executive producer. We listened and created assets for those who are "below the line" to communicate how important it is to replace plastic with reusables. We want to show that this shift away from single-use plastic is being requested by the heart of the industry, as well as at the request of environmental organizations like ours.

We were selected to be a part of the Screen Actors Guild "Green Council," which was born under the care and consideration of the guild's current president, Fran Drescher. She was able to create a coalition of nonprofit organizations and for-profit groups to come together and eliminate all single-use plastic from every set throughout Hollywood. We are very proud to be a part of this and happily attended the

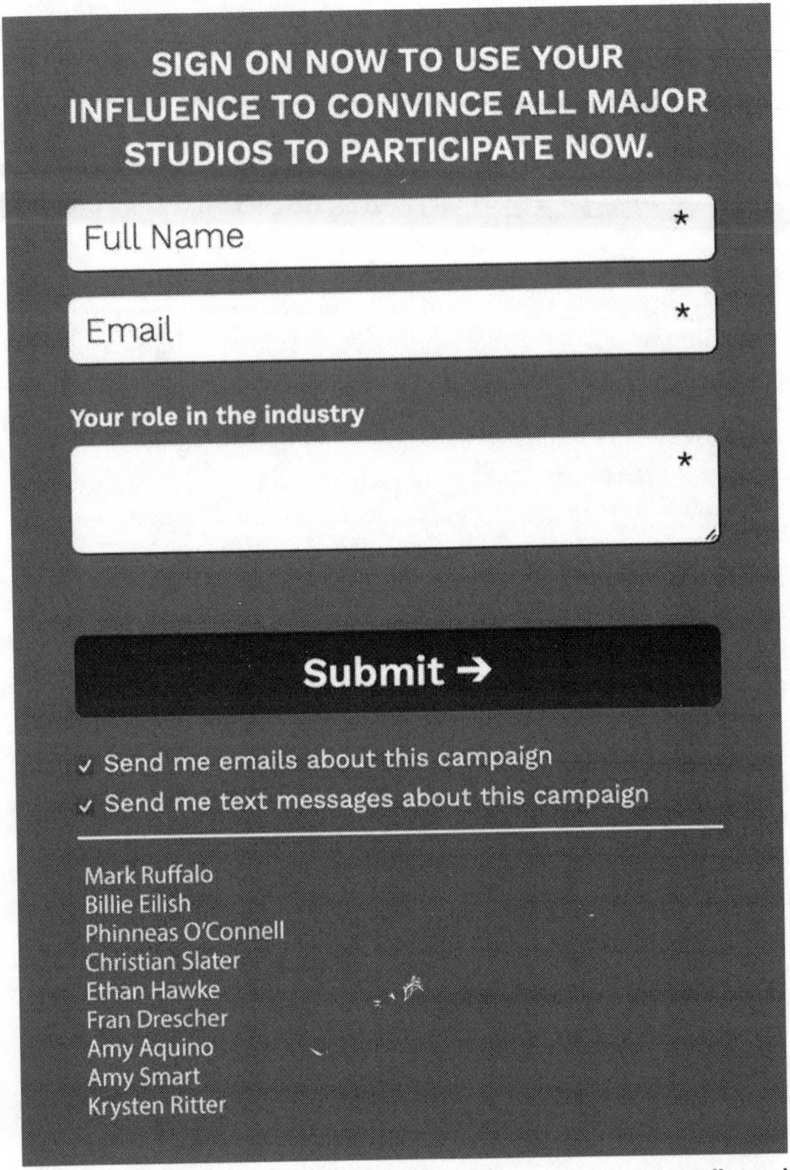

SIGN ON NOW TO USE YOUR INFLUENCE TO CONVINCE ALL MAJOR STUDIOS TO PARTICIPATE NOW.

Full Name *

Email *

Your role in the industry *

Submit →

✓ Send me emails about this campaign

✓ Send me text messages about this campaign

Mark Ruffalo
Billie Eilish
Phinneas O'Connell
Christian Slater
Ethan Hawke
Fran Drescher
Amy Aquino
Amy Smart
Krysten Ritter

Figure 7.2. Petition by Habits of Waste to reduce plastic use in Hollywood signed by Billie Eilish, Finneas, Christian Slater, Ethan Hawke, Fran Drescher, and many others. Image courtesy of Habits of Waste.

inaugural event with the Motion Picture Association in Washington, D.C., to announce the Green Council and the commitment to end single-use plastic in Hollywood.

As our campaign expanded, we started working within the growing movement for climate storytelling, which entails everything about bringing climate change into the storylines in films and television shows. We love hearing about these big goals and fully support climate storytelling. The #LightsCameraPlastic? campaign is considered the first step toward these bigger goals, as it is the perfect low-hanging fruit to get people started on doing something positive (some say it's the gateway to more, but it is an entry point that virtually any film or television production can get behind). In order to start building a new societal norm—that reusables are preferred or cooler than single-use plastic—we must take small, achievable steps. Some productions may not be ready to change their scripts yet, but swapping single-use plastic for reusables in scenes where it makes no difference at all is very doable for most. This is a grassroots force that the fossil fuel industry cannot fight no matter how much money the companies pour into misleading the public about plastic.

We are seeing huge progress. Universal Pictures just launched a massive green initiative called the GreenerLight Sustainability Movement. Degen Pener from the *Hollywood Reporter* wrote about this initiative in a piece that references our work as well: "The studio also said the program will encompass looking at depictions of on-screen behaviors from an environmentally friendly angle. (Similar moves have been made by such directors as Kat Coiro, who has worked in partnership with the nonprofit Habits of Waste to show characters using reusable containers instead of single-use plastics onscreen.)"[19]

The plastic-pollution crisis is so massive that we really must get creative and go at it from all angles. If our favorite actors and actresses are using single-use plastic on-screen, then viewers believe it is OK for them to use it too. It's amazing to see how impressionable we are. When I spoke to the world's number-one story scientist, Dr. Angus Fletcher, he noted that when we see reusables on-screen, our minds open to new ideas and options.[20] We are then able to see the solutions and perhaps be inspired to make the change within our own households.

We aren't done yet. We have much work left to do in the entertainment industry, but we have already expanded the movement to

Figure 7.3. Flyer for the #LightsCameraPlastic? campaign by Habits of Waste. Image courtesy of Habits of Waste.

the sports industry, as well. We launched #LightsCameraPlastic? Sports Edition in 2023, which continues the effort in a similar manner by pushing athletes and their teams to rehydrate with reusable alternatives. This means that we must convince the major sports leagues such as the NBA, the MLB, the NHL, the NFL, soccer and football organizations worldwide, and even individual sports like tennis to also participate. The sports industry is also responsible for setting the tone for what is and is not socially acceptable. Plus, we are trying to help athletes learn from studies that show how drinking water out of a single-use plastic water bottle is not healthy for our bodies due to the microplastics that are released from the plastic water bottle itself.[21] We invite you to join the #LightsCameraPlastic? movement and participate in our call to action targeting the sports industry on every level. We have already seen Wimbledon 2023 provide athletes with reusable water bottles.

HOW YOU CAN BE AN IMPERFECT ENVIRONMENTALIST

1. If you see single-use plastic in a television show or film you are watching, take a photo and post it. Tag the show and ask them to try reusables next time. It's a great way to keep putting the pressure on the studios and dictating what we do and don't want. If anyone asks you why this matters, remind them that corporations spend millions of dollars on product placement ads because *they work*. Companies understand that people will be impacted when their products are shown on television and film.
2. Sign on to our #LightsCameraPlastic? call-to-action email to help us put pressure on the film and television industry to begin denormalizing single-use plastic by removing it from our films and television shows.
3. Use social media to promote change. If you see a celebrity or an athlete pictured with single-use plastic, let them know that this is a bad way forward and that they can use their huge platform for good.
4. Sign on to our #LightsCameraPlastic? Sports Edition call-to-action email to help us put pressure on the sports industry to

begin denormalizing single-use plastic by removing it from the sidelines.

5. Swap for reusables any chance you get in your everyday life. Do you know about the chameleon effect? We tend to unconsciously imitate the people around us in a phenomenon researchers have termed the chameleon effect. Let's set a great example by showing off our reusables next time we are working out at the gym, going to the grocery store, or grabbing coffee.
6. Tag celebrities and ask them to sign our petition so that we can continue growing our list with strong names to convince studios to adopt #LightsCameraPlastic?
7. Talk about this idea with your friends. The more people know about solutions like #LightsCameraPlastic? the more they can help create the momentum needed for lasting change.
8. Show off your reusables so we can influence Hollywood to see that reusables are trending and here to stay. After all, Hollywood is trying to emulate our lives on-screen. Life imitates art. Let's remember to imitate sustainably.
9. The entertainment industry is powerful, but it's made up of a small group of people. We can convince them to make the necessary changes. Check your network and see if you can spread the word inside Hollywood.
10. Films and television shows deserve an eco-rating. Tag platforms like @rottentomatoes to consider this rating system.

8

#8MEALS

If there was ever a campaign that welcomed imperfection, this one takes the cake—vegan cake, that is. It all began because I felt like a massive failure. From as far back as I can remember, I recall resisting anything that contained meat or eggs and even butter. The first memory I have of this was my mom forcing me to eat a soft-boiled egg for breakfast. Just the thought of this egg still makes me cringe. She made me eat it because she thought I needed it to be an attentive student at school. In order to appease her and get to school on time, I forced myself to swallow the slimy and very watery egg, which I ended up vomiting on the sidewalk before getting into the car to go to school.

This episode did not change the routine for us that much, but my mom did agree to give me hard-boiled eggs instead. It was better but still very hard for me to swallow, literally and figuratively. As time went on and I entered my teen and adult years, I noticed that I left the chicken from my chicken salad on my plate at the end of my meal, but I felt that I "needed" to order the chicken salad because of the belief system that was instilled in me from a young age that we humans needed animal protein to survive. I rarely ever wanted anything to do with animal products at all but would continue to order food with meat because of my upbringing that a good nutrition revolved around "protein," animal protein that is. I was taught that those who didn't eat enough protein would go bald, have no muscles, be weak and sleepy all the time, and were just basically unhealthy. How far this all was from the truth, I really didn't know. So I kept ordering "protein" despite disliking it. I discovered that the only way I'd be able to eat it was if it were cooked well or even overdone. This had its own challenges in finer restaurants because it's considered a major faux pas to eat this way.

One day I was at lunch with a friend and our small kids. I ordered a salad with some salmon on it. My friend, who was vegetarian, ordered an eggplant dish called baba ghanoush that came with chopped walnuts, caramelized onions, and little pomegranate seeds on top, and was served with warm, fluffy pita bread. She generously offered me one pita wedge to try with the dip. To this day I can remember the taste of that heavenly bite. I wished I had ordered the same thing she had. The next time we went to brunch she ordered lemon ricotta pancakes with slivered almonds inside the batter. I had ordered scrambled eggs, which were runny and just plain gross. As I tried a bite of her pancake, I remember feeling the crunch of the almonds with the lemony flavor of the pancake, and it was sublime.

Practically each and every time I dined with this friend of mine, I wanted her food instead of my protein-filled meal. I decided then, once again, that I would commit to going vegan (I had attempted this goal several times by this point in my life). It lasted about two weeks. I went to my mom's for dinner and she made another one of her amazing stews. Persian stews are usually made with vegetables, herbs, onions, and some kind of meat like lamb or veal shank. I'd find myself eating everything except the meat that was hiding inside. I had already told my mom that I was trying to commit to veganism; she replied that I could just have the juices of the stew, not the meat. I was reminded of the scene in the film *My Big Fat Greek Wedding* where the main character, Ian Miller, tells his fiancée's family that he doesn't eat meat, and they said no problem, we have lamb! I have learned that there are myriad ways to make Persian stews vegan; jackfruit or mushrooms are my preferred meat replacements.

Needless to say, my numerous attempts to go vegan went much like that two-week stint. I would try so hard and then come across a cheese platter that was practically calling my name. Or I would want to have frozen yogurt or something that was not fully plant based, and each and every time I caved into my nonvegan cravings, I'd feel like I had failed yet again and fallen off the imaginary vegan band wagon.

When did food start to make us feel like failures? Since when do we attach labels to ourselves that announce what kind of food we permit ourselves to eat? I am sure you have heard someone you know state that they are "paleo" or "keto" or "gluten free," "low carb," or "pescatarian," "vegan," or "vegetarian," and so on. There is so much attached to these labels, including stereotyping and even discrimination. What if we all just focused on doing our best, one meal at a time?

In August 2020 I was looking out the window during a family road trip to my happy place, beautiful Lake Tahoe, California, which I go to every summer to immerse myself in nature. We were on Interstate 5, and it was a sunny and very hot August day. The temperature was close to 110 degrees Fahrenheit as we inched our way closer to Sacramento. As I looked out the window I noticed huge cattle ranches with thousands of cows standing around on dirt (not grass) with zero shade and zero water anywhere in sight. The stench that comes from these ranches is virtually unbearable for several miles. We continued our drive north and passed another farm, but this time we saw hundreds of calves caged up separately from the adult cows. Unlike the older cows, which had space to move, these babies were kept inside small plastic containers, which were surely many degrees hotter than the temperature outside, as there was no shade anywhere to be found here, either. Their little plastic container had a cage at the opening, which allowed them to take one step outside to get water, and then they had to go reverse step and go back inside of their little ovens since there wasn't even enough room to turn around. Someone explained that the calves are eventually slaughtered and sold as veal. The reason they don't have much room to move around is because they aren't supposed to move. This forced immobility is done intentionally so that their muscles don't become tough for us humans to eat. Seeing their confinement in that heat brought me to tears.

I remember turning to my husband and saying, "How can we humans be so cruel?" How does the rancher sleep at night knowing there is so much suffering happening outside under his watch? It isn't just the rancher that I think of. Equally as mind-blowing is how humans can disconnect so fully from the cruelty that goes into getting that meat prepared from beginning to end. It's as though the meat just magically ends up on their plates with no impact whatsoever. It's as though we humans deserve to have whatever we want without any regard for other living beings on this planet because, after all, we are "superior." Ironically, what we are doing is hurting ourselves in the long run.

I have heard from many who gave up meat and only eat fish. Being a pescatarian has been on the rise because of the bad rap other animal products have been facing in recent years. In fact, the biggest threat to our oceans today is overfishing, which depletes species' breeding populations to a point from which they cannot recover. This destabilizes ecosystems and threatens the supply of fish for consumption.[1]

Figure 8.1. The 8meals app, launched by Habits of Waste in 2021, offers meal suggestions and tracking options with an "imperfect environmentalist" spin. Image courtesy of Habits of Waste.

All of these pieces came together and became the straw that broke the camel's back. I decided that enough was enough and I needed to take action. As I racked my brain for answers, I realized that it was all very simple. We all needed to eat more plant-based food. Period. However, from growing up in my household, I already knew that this was virtually impossible. If I struggled with veganism for so long, then surely others who had attempted it felt similarly. I was well aware that many people never even think about going vegan and do not consider or understand the connection to the environment or the suffering of the animals. Only 3 percent of the population was actually living a vegan lifestyle, which meant we had 97 percent of the population to work with on making change.[2]

Ninety-seven percent. That was a huge number of people to impact, and that gave me hope. So I began researching to find answers, any answers. Instinctively, I knew that showing the animal cruelty happening in factory farming wasn't the answer. I have seen the most painful and horrific videos, which are truly gut wrenching, but I believe they shut people down from getting involved because it is just too much for them to handle. If we are lucky, those videos will make a small percentage of people go vegan. But I wanted to change the norm around eating animal products altogether, which meant we needed to bring people in by shining a light on the positive. I wanted to help provide that large group of nonvegans with an opportunity to do better without shaming or blaming them.

I studied everything I could on the connection between our food and climate change. I searched for weeks for something that I could run with to start making drastic change. One day I finally discovered a study called "Implications of Future U.S. Diet Scenarios on Greenhouse Gas Emissions."[3] It was written by Dr. Martin Heller and Dr. Gregory Keoleian, both from the University of Michigan, and Dr. Diego Rose from Tulane University. Their thirty-page study proved that if we commit to *reducing* our animal protein intake, then we can significantly begin to combat climate change. I felt as though I had just won the lottery. We didn't need everyone to go fully vegan, but we did need everyone to decrease their consumption of animal products. I realized that this study is a gem that most individuals would probably never see, so I provided a link to the study on our website at www.habitsofwaste

.org/campaigns/8meals/ so everyone can see for themselves that science exists proving that a partially plant-based diet does have significant value toward combating climate change.

I was very focused on the calculations in the study clearly stating that a decrease of 40–50 percent in animal products would create a huge carbon offset that could collectively combat climate change altogether. I calculated that an average person eats about 21 meals per week (3 meals per day × 7 days a week = 21 meals), then I subtracted 40 percent of those meals, which left us with 8 meals that needed to be plant-based in order to make a real impact. With this information I created a new campaign to help people set a reasonable goal each week, without a polarizing label that would inevitably leave them feeling like a failure. That is when #8meals was born.

Meeting one of the scientists behind the study was essential because I wanted to express my gratitude for their amazing work and to share how inspired I was that it didn't have to be "all or nothing"—or better yet, vegan or nonvegan—to make a difference. In fact, I was just so excited to remove those labels from our collective mindset and start to create a new ideology altogether, one that fits in with the concept of being an imperfect environmentalist.

Setting up a call with Dr. Heller was my next priority, and to my excitement he replied to my email requesting some time with him. When we meet via Zoom, I explained that I was very intrigued by his work and wanted to translate the study into something that the mainstream population could use as a guide toward reducing their carbon footprint while halting the growth of the meat and cattle industry and the horrific factory farming of animals.

He was so happy to hear that I had discovered his study, and he was pleased that I was trying to create a change-making campaign based on his work. If there was ever a moment that I needed a little boost to get me going, Dr. Heller provided it for me. He even helped me with an incredible calculation to show the impact of #8meals. If we all tried to decrease the animal protein intake for ourselves and our families by 40 percent each week, it would be the equivalent of driving a hybrid car for one year. Dr. Heller said it best: "We cannot wait for governments and corporations to act. Perhaps the single greatest contribution we can make as individuals is to change our diet. The carbon footprint reduction of committing to at least 8 plant-based meals a week is on par with switching to a hybrid automobile for the average American family."[4]

It's rare to compare what you eat to the type of car you drive, but this was a great concept for me to use to show people that they could really help the planet. But I was still trying to find the win-win-win that had made my other campaigns so successful. As I started to build the campaign, my first step was to recognize how food is so personal and why people have so many feelings around it. It's very different from the work I had done to ban single-use plastic or save crayons. Here's how personal it really is. One study analyzing restaurant menus shows that when a restaurant menu provides vegan options mixed in with nonvegan dishes, the chances of people ordering a vegan meal is much higher. On the contrary, when they list the vegan options in their own section, people don't order these items as much because they feel they must consider themselves vegan or vegetarian to order from that section.[5] These labels are another barrier to cross, and I think that is a major issue we must change throughout society. Thinking back on the meals I shared with my vegetarian friend, I am reminded that I too was guilty of this behavior. Even I didn't order "vegetarian" food because I wasn't technically a "vegetarian," though I always loved the food my vegetarian friend selected more than my animal-protein-heavy meals.

Another important step to build this movement was to start renaming the food we eat as "plant based" versus "vegan." This was a genius move by whoever began this "plant-based" movement because the word "vegan" brings up so many stereotypes and holds a lot of stigma. Unfortunately, many times vegans are depicted on television as extreme environmentalists who are totally out of touch with mainstream society. Some say that being vegan is also a philosophy or way of living complete with deep value systems. Much of environmentalism is about marketing to the masses, and rewording things to make them feel more inviting and inclusive is always helpful, so I made a pact to refer to the #8meals campaign as plant-based versus vegan. To this day there is less stigma when a person says they eat a plant-based diet, and we hope it stays this way.

Next we created the "#8meals Challenge" on our website, complete with a sample week that highlighted eight of your meals as suggested plant-based meals. For example, one week could have two breakfasts, three lunches, and three dinners that would all be plant based.

Another week could have one breakfast, two lunches, and five dinners that are all plant based. It seemed to give people a method to go about this challenge, and we felt that the best time to start promoting the campaign would be right after Veganuary on February 1, 2021. Veganuary is a global challenge whereby participants go vegan for the entire month of January.[6] We wanted to catch all those people coming off Veganuary so that they didn't go back to their regular diets; rather, we wanted to introduce them to our "partially plant-based" campaign, which we hoped would become a yearlong commitment.

I knew that I had something great because the response was astounding. People were so relieved to know that they could make an impact without going all or nothing. Our media partners and several plant-based influencers worked on helping us share the campaign far and wide. While the excitement was palpable, a common concern was that people wanted someplace to "go" to get started. I cannot tell you how many times I heard the question, "Where do I go to start?" I realized that this concept of taking the leap to a plant-based diet in any capacity was a big one for most and that they needed more support and guidance. One of my friends suggested I create an app that would be free, like a "plant-based diet for beginners" tool. It would surely help everyone get started.

As luck would have it, a friend in the tech world, who happened to be a parent in my son's elementary school, learned about my work and offered to help me anytime I had any questions or needed an app. So I took my chances and took him up on his offer. He introduced me to the founder of an app company called Digital Pomegranate. This man, Todd, was like an angel to us. Within about three months I had an app called "Habits of Waste" with the tagline "Plant-based starts here." The notion of the app is that there are zero expectations—just give the #8meals Challenge a try.

The app helps users find whole-food, plant-based recipes, plugs them into the calendar on their phones, and provides reminders. It also tracks your carbon offset each week so that you can see your progress and how much you are impacting the planet. We wanted to make the challenge fun and light, so we gamified it and included stickers people could add to their photos and then upload to social media. Users could also share recipes with friends, calculate their carbon offset, and check where their friends are. They also received reminders to keep going to reach their #8meals goal. Most important, we added a quiz to educate people and dispel the many misconceptions about eating a plant-based diet. Education is key for more people to become less intimidated and give it a try.

We decided to only go with whole-food, plant-based recipes to ensure we were including everyone in all socioeconomic positions. The term "whole food" refers to food that is natural and minimally processed.[7] Meat and dairy alternatives tend to be very expensive, so we avoid those in order to dispel the myth that eating a plant-based diet is very expensive. In fact, we did some calculations on a few of our recipes and found that you can feed a family of four for under $10 since the main sources of protein in our recipes are budget-friendly options such as beans. One recipe that I love is a vegan lentil curry with rice. It's nutritious, delicious, satisfying, and very reasonably priced.

Tech Crunch wrote a wonderful article about us, which increased our downloads right off the bat.[8] Some people thought we had spent at least $100,000 on this app, but thanks to Todd and his team, we were blessed with the most wonderful group of developers, who created the app almost pro bono because Todd believed in us and our idea that we can support people through technology in their plant-based journey. One of the lead developers was vegan, so she worked with all her heart to build us the most beautiful and user-friendly app. She was happy to support this approachable campaign to help more people successfully get involved in the vegan movement that was starting to take the world by storm. All of this couldn't have come at a better time.

That summer I learned that the "Lungs of the World" were on fire. In August 2021 the Amazon rainforest had one of the largest fires in history. Everyone was devastated. As part of the response to this catastrophe, I began hearing a connection being made between the Amazon fires and the hamburgers we eat. Being from California, I was used to fires—sadly, a little too used to them. Nonetheless, we blamed a multitude of things for our California fires, but hamburgers were never one of them.

Here is the connection. The majority of horrific Amazon rainforest fires were not accidental. In fact, they were intentional. The reason for this: Brazilian loggers and ranchers wanted to clear the land for cattle,

Figure 8.2. The 8meals app. Image courtesy of Habits of Waste.

which the Brazilian populist pro-business president permitted, as he was backed by the country's "beef caucus."[9]

Seventy-two percent of the major fires in 2022 occurred in the Brazilian Amazon, and of those, 71 percent were caused by in an effort to deforest the area.[10] It doesn't stop here. The main cause of deforestation in almost every Amazon country is cattle ranching, which currently accounts for 80 percent of deforestation.[11] To make matters worse, all these new cattle ranches need to feed their cows in order to "fatten them up," which means more land is needed to farm corn and grains. So the dreaded cycle continues: more deforestation to grow the feed needed for the increasing number of cattle raised.

Beyond forest conversion, cattle pastures increase the risk of fire and are a significant degrader of riparian areas, the land along the edges of rivers, streams, lakes, and other water bodies. Cattle pastures are a major disruptor of aquatic ecosystems, causing soil erosion, river siltation, and contamination with organic matter. Turning forests into pastures that are then exposed to overgrazing leads to extreme loss of topsoil and organic matter, which may take decades or centuries to replace. Unfortunately, trends show that livestock production is expanding in the Amazon.[12]

There's more, and it gets worse. Three hundred forty million tons of carbon—equivalent to 3.4 percent of current global emissions—are released into the atmosphere every year from farming cows.[13] Then there is the emission of methane, another greenhouse gas that traps heat in the atmosphere. Essentially, when a cow burps, methane is emitted into the atmosphere. Cow belching results from enteric fermentation, when digestion converts sugars into simple molecules to be absorbed into the bloodstream, which produces methane as a by-product.[14] Additionally, methane is also emitted from the decomposition of cow manure under certain conditions.[15] Researchers estimate that the cattle industry is responsible for 14.5 percent of global greenhouse gas emissions.

Many wonder how we can quantify the methane produced in cow burps. At the University of California, Davis, researchers provided this data by placing a Holstein cow with its head and neck sealed airtight inside a large, clear-plastic chamber that resembles an incubator for newborns. While giant tubes above the chamber pump air in and push air out, the cow calmly stands and eats her feed. Equipment inside a nearby trailer spits out the data needed to measure the methane output. We need to learn this information in order to inform naysayers how much the cattle industry is ruining our planet.[16]

Needless to say, I was so happy that we had found one solution through our plant-based campaign. We were turning even the most unlikely people into plant-based enthusiasts. The testimonials that came to us after a few months were surprising, to say the least. We heard from many that they had never thought they could go plant based at all, but by having the pressure taken off them to be fully vegan, they easily committed to eight meals and then slowly began to increase to nine meals, then ten, and so on. One participant, ironically, was the son of a butcher in Ireland. He took the #8meals Challenge and within a few short months committed to going fully vegan. The best part of his story is that he was on many medications due to what he thought was a stomach illness. He not only stopped needing medication for his stomach issues but

Figure 8.3. Preparing examples of plant-based meals to promote the new app virtually. Photograph by Sheila Morovati.

also lost over forty pounds. Doctors attributed this to his new habit of eating eight plant-based meals a week, which is excellent for heart health and very beneficial for overall health, in addition to really doing a lot to help protect the planet. In fact, Columbia University Irving Medical Center states on their website that "a plant-based diet is considered to be nutrient-dense and packed with fiber, healthy fats, protein, vitamins, and minerals. It is a very healthy way of eating and can meet all of your nutrient needs."[17]

This app and the #8meals campaign led to an invitation to join the World Economic Forum's Nutrition Disruptors, a group of experts who are invited to come together in an effort to solve large-scale problems. I sat at a virtual roundtable with huge corporations like Tyson Foods, PepsiCo, Kroger, Mars, Inc., and more. We discussed the way to a healthier future, and plant-based eating was at the top of the list. In fact, Tyson Foods, which is one of the world's largest producers of poultry, pork and beef, has since come out with a line of plant-based alternatives. Trends are that "the global plant-based meat market size was valued at USD 4.40 billion in 2022 and is expected to grow at a compound annual growth rate (CAGR) of 24.9% from 2023 to 2030."[18] This makes us feel very hopeful, but we must continue helping more people commit to eating a plant-based diet.

We have also been included in the World Economic Forum's Consumers for Climate Action and their Global Future Council for Net Zero Living, which is cochaired by the chief sustainability officer for Google and the chair of sociology at UCLA. Being a part of this group of world leaders who come together to find solutions for a cleaner future is an honor that I cherish.

We are still working hard to bring #8meals to the masses. One way we do this is by meeting with large corporations that are looking for ways to reach their environmental, social, and governance goals by offsetting carbon or being carbon neutral. We have been able to show them that their team members can take the #8meals Challenge and collect the data on all their employees' progress in order to calculate their collective carbon offset. Gamification is a great way to engage team members, or even ask a group of friends to join in and do it together. It's simple and easy enough to try without any chance of "failing" because even one plant-based meal impacts the planet.

Imagine the ripple effect that would ensue if we all started to eat just one plant-based meal. As I mentioned in previous chapters, large corporations are watching us take the lead, basing their next move on the direction we are taking. So let's keep upping those plant-based meals and letting everyone know that we are ready to give veganism a try. Remember, eating a plant-based diet is the most important step any individual can take to protect our planet.[19]

HOW YOU CAN BE AN IMPERFECT ENVIRONMENTALIST

1. Go plant-based if you can. If not, start with a goal of being partially plant-based by committing to eight plant-based meals each week. Ask a friend to join you in reaching this goal!

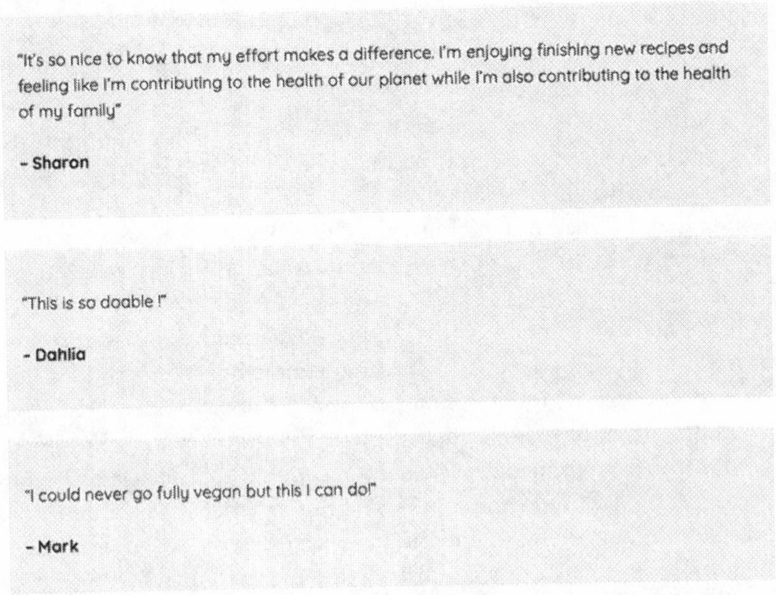

"It's so nice to know that my effort makes a difference. I'm enjoying finishing new recipes and feeling like I'm contributing to the health of our planet while I'm also contributing to the health of my family"

- Sharon

"This is so doable !"

- Dahlia

"I could never go fully vegan but this I can do!"

- Mark

Figure 8.4. Testimonials from app users. Image courtesy of Habits of Waste.

2. Download our app from the App Store or in Google Play (under "Habits of Waste"). It's free and there to help you reach your plant-based goals.
3. If you are going to continue eating meat, fish, or chicken, consider buying locally from a humane, certified farmer.
4. When dining out, order the plant-based options available to you and make a request to the chef to offer even more plant-based options on their menu.
5. Share photos of your amazing plant-based meals on social media. People are often unaware of how delicious and filling vegan meals can be. We must dispel the myth that vegan meals are just green salads!
6. Turn vegetarian recipes into vegan ones easily by swapping for plant-based ingredients. For example, you make a recipe fully plant based with little effort and likely not even notice a difference in flavor by switching from butter to olive oil or avocado oil.
7. Many staples we love are already plant based. Give new foods a try too. For example, Indian food is a great way to eat a plant-based meal and feel truly satiated. The same is true for many Italian pasta dishes. Of course, good old guacamole and chips make a wonderful plant-based appetizer.
8. Ask your local supermarket for more plant-based options.
9. Start small. Try putting a nondairy creamer in your coffee for a few days. Chances are that you will end up liking it and continue using it.
10. Be open-minded. Put all those myths aside. Going plant based is not a punishment! It's actually a great way to try something new and see how great you feel. Listen to your body and give it a chance. You might love it and never look back.

9

CARBON OFFSET CHALLENGE AND HOW EVERY ACTION HAS A VALUE

This chapter is the "how-to" guide to make changes in your home and at work, which will launch your transition to becoming an imperfect environmentalist. I will go through everyday habits that you can easily replace with more sustainable alternatives that will truly impact the state of our planet. I have a few ideas to share that are not difficult to adopt but will be game changing for the health of our planet.

Part of my work is to analyze the "habits of waste" in my daily routine and the daily routines of my family members. While doing this I began to observe daily life through a lens of sustainability from the time I wake up to the time I go to bed. With each discovery I noticed that there is more I could do, which tells me that there is more everyone can do. As I focused on daily habits, I simultaneously looked for alternative opportunities that were more eco-friendly. If there's anything I'd like for readers to take away from this book, it is that we can create huge ripple effects if we just start somewhere.

The words "carbon offset," "net zero," and "carbon neutral" were popping up at every turn, and I wondered, "Does anyone even understand all this terminology being thrown around?" There is so much talk about carbon that is derived from emissions or industrial agriculture, but not so much from our individual lives. It seemed fitting to establish our carbon output so we could quantify the change we are asking people to make and create a connection to understand where this invisible gas comes from. Showing the levels of carbon that come with each one of our daily actions provides proof that we can do better and make a difference. The goal is for the world to have more context so that over time we can all become less wasteful.

You may have heard people say that we as individuals don't have any power and cannot save our planet. The onus of climate change shouldn't be on us. In many ways I do agree with this. However, what we *can* do is create a groundswell, which becomes a grassroots movement that can have so much momentum that it makes change within corporations and in legislation. An organized group of people with a purpose and goal in mind to create change has historically been the beginning of so many successful social movements (e.g., the civil rights movement and women's suffrage). So, as an imperative step, I wanted to create something to empower individuals to take action versus feeling so paralyzed by climate anxiety that they feel too small to do anything. The manageable steps provided in this chapter are meant to help individuals achieve small wins each day, consequently building their own internal power to start conquering the huge problems created by greed, such as those fueled by the fossil fuel industry. We must do this so that future generations may hope to have a habitable planet.

My team and I decided to create a game called "The Carbon Offset Challenge," which is available for free at www.habitsofwaste.org. The interesting thing about this challenge is how we place a carbon value on our daily actions (see photo below). Each week offers you a chance to do better and generate less carbon through small shifts in your habits. The calculations are all in the chart, so you don't have to do much except fill in your actions. Here's how it works:

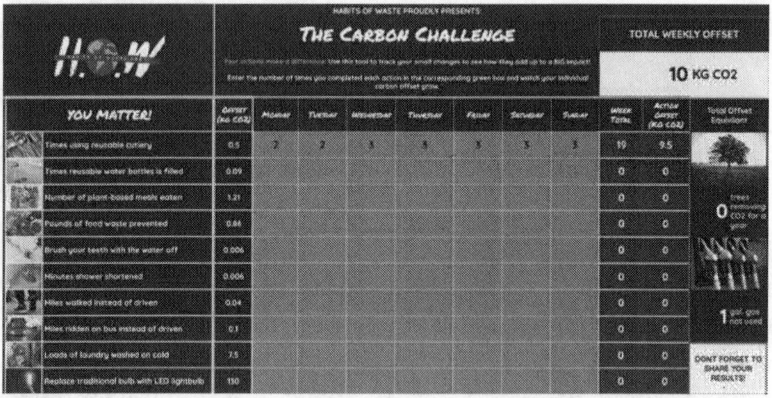

HABITS OF WASTE PROUDLY PRESENTS:

THE CARBON CHALLENGE

TOTAL WEEKLY OFFSET: 10 KG CO2

YOU MATTER!	Offset (KG CO2)	Monday	Tuesday	Wednesday	Thursday	Friday	Saturday	Sunday	Week Total	Action Offset (KG CO2)
Times using reusable cutlery	0.5	2	2	3	3	3	3	3	19	9.5
Times reusable water bottles is filled	0.09								0	0
Number of plant-based meals eaten	1.21								0	0
Pounds of food waste prevented	0.84								0	0
Brush your teeth with the water off	0.006								0	0
Minutes shower shortened	0.006								0	0
Miles walked instead of driven	0.04								0	0
Miles ridden on bus instead of driven	0.1								0	0
Loads of laundry washed on cold	7.5								0	0
Replace traditional bulb with LED lightbulb	150								0	0

DONT FORGET TO SHARE YOUR RESULTS!

Figure 9.1. The Habits of Waste Carbon Challenge, which helps individuals track the positive environmental impact of small changes they have made in their daily lives. Image courtesy of Habits of Waste.

STEP 1: Register for and then sign into your account to save your progress and calculate your cumulative carbon offset over time.
STEP 2: Review the list of actions and make small, thoughtful swaps to your habits of waste each day.
STEP 3: Enter your daily totals into the dashboard below to calculate your cumulative weekly impact.
STEP 4: Download your impact graphic, share it on social media, and challenge your friends to see who can make the biggest impact!

I used the small shifts I made in my own behavior as inspirational starting points to help launch the Carbon Offset Challenge, which applies a carbon value to any action an individual takes on any given day. This offset value can motivate people to achieve behavior change—you can significantly decrease your carbon footprint by simply changing a small habit of waste. This online calculator gives you tangible data about your actions each day. At the end of the seventh day you receive a score, which is the amount of carbon you were able to eliminate from the atmosphere.

The key is to get the masses on board so they too can understand the cause and effect of eliminating carbon from their lives by making a series of small, attainable changes. After all, close to eight billion people live on this planet, and we must all assume the responsibility of our actions, especially as they pertain to the planet around us. Starting small is the secret to long-term success. My story is the perfect example of this. I began my environmental journey simply by trying to salvage restaurant crayons. My sole focus for a period of time was just the crayons, but then it blossomed into so much more. When it came to adopting a vegan diet, I started by switching the dairy creamer to almond milk creamer in my morning cup of coffee and built from there.

As this experiment progressed, I realized that we could do better in many ways, but we must try to "see" things around us more clearly, which is what this chapter is going to help you with. The following is a thorough list of ways to reduce some habits of waste with minimal effort so you can really help the planet. At times you will be surprised to know that you can also save money while doing this.

Save Energy and Money

1. Wash your clothes in cold water and use a clothesline or drying rack to dry your clothes. Why? On average, American households do about three hundred loads of laundry every year. Heating the water and running the machines consumes enormous amounts of energy, which translates into more greenhouse gas emissions.

   "One study estimated the nation's residential laundry carbon dioxide emissions at 179 million metric tons per year. That's equal to the total annual energy use of more than 21 million homes."[1]
2. Save energy by turning off lights and using natural light or switching to LED light bulbs. According to the U.S. Department of Energy, compared to incandescent bulbs, residential LED light bulbs use at least 75 percent less energy and last twenty-five times longer.[2] Not only does this mean a lower electric bill, but also fewer times you'll need to replace the bulbs.
3. Keep your thermostat at 70–78 degrees Fahrenheit (°F). Energy Star, a joint federal program run by the Department of Energy and the Environmental Protection Agency, recommends that for optimal cooling and energy efficiency, the coolest you should keep your house is 78°F—and that's only when you're at home and awake. A smart or programmable thermostat makes it easy to match your cooling needs to your schedule, but you can make the adjustments manually if you don't have one for your central air system. Try the following settings:

   - 78°F when you're home
   - 85°F when you're at work or away
   - 82°F when you're sleeping

4. If you're frustrated with high utility bills, you might be interested in switching to green energy such as solar power. With solar panels, you can generate power yourself, reducing energy costs and your reliance on the public grid. Many cities also offer incentives so more people can install solar panels in place of

traditional energy sources, providing clean power all year long for their homes, businesses, or vehicles.

5. If you can switch to an electric car, great; if not, consider a small, fuel-efficient car. Try to walk or bike to places from time to time. Take public transportation or carpool whenever possible. Each mile we drive equals four hundred grams of carbon dioxide, which adds up to an average of 4.6 metric tons of carbon dioxide per vehicle per year.[3]

Go Reusable

Single-use products are so damaging to our planet, especially when you think about the mass quantities we use worldwide.

1. Use reusable coffee/tea mugs or bring your tumbler to your local coffee shop. Five hundred billion disposable coffee cups are used globally each year. These cups are lined with plastic, so the liquid doesn't leak out, and take more than twenty years to break down.[4]
2. If you make your coffee at home, consider using a refillable pod. They're less expensive, and their use will reduce the sixty-two billion single-use coffee pods consumed annually in North America and Europe.[5]
3. BYO (bring your own) reusables! As mentioned earlier, five hundred million plastic straws are discarded per day just in the United States, and forty billion pieces of plastic cutlery are produced per year only to be used once and thrown away. Keep some reusable cutlery and straws at work or in your car for easy access.
4. Use a reusable water bottle. The most important item I never leave home without is my reusable water bottle. My office has filtered tap water, and I challenge myself to refill my bottle multiple times a day. It definitely beats the cost and the environmental damage of the nearly six hundred billion plastic bottles and containers we discard each year, which equates to around twenty-five million tons of plastic waste—most of which is not recycled and ends up in landfills.[6]

5. Take reusable bags when you go grocery shopping so we can decrease the five trillion plastic bags produced worldwide annually. In fact, I leave a small array of reusables in the trunk of my car so that I have anything I need in case I want to take leftovers home from a restaurant, I need grocery bags or a coffee cup, and so forth.
6. Another neat trick is to save glass jars to use as (or instead of) "Tupperware." I have been known to have many salads in old pickle jars or fill jars with homemade juices and smoothies. Remember, you don't have to spend a lot of money to be sustainable.
7. Get rid of all the plastic wrap in your kitchen. Try reusable alternatives like beeswax covers or silicone lids that you can place over any dish or platter to keep the food fresh. Did you know that each person uses 38.14 pounds of plastic film per year?[7]
8. Say goodbye to plastic baggies. They are a massive polluter and cost you a lot of money. American families use five hundred Ziploc plastic bags a year, which means as a nation we are using forty-two billion plastic bags a year.[8] We are paying millions of dollars for single-use plastic baggies that can be replaced with a reusable alternative that we would have to purchase once instead of constantly replenishing.

Food Waste and Composting

Food waste is one of the worst culprits of methane gas emissions, and the United States discards 50 percent of the food produced.[9] This is saddening on many levels, especially because so many people face food insecurity in the United States and beyond our borders. The U.S. Environmental Protection Agency (EPA) published a report in 2021 on the environmental impacts of food waste, estimating that "each year U.S. food loss and waste embodies 170 million $MTCO_{2e}$ GHG emissions (excluding landfill emissions)—equal to the annual CO_2 emissions of 42 coal-fired power plants." This estimate does not even include the methane emissions from food waste rotting in landfills. EPA data show that "food waste is the single most common material landfilled and incinerated in the United States, comprising 24 and 22 percent of landfilled and combusted municipal solid waste, respectively."[10]

1. Instead of throwing food away, you can compost it to create rich, nutrient-dense soil. Some cities, like Los Angeles, will do the composting for you, so you don't need to handle worms and all the dirty work; simply place your produce scraps in the green bin. You can reduce your family's methane output significantly. Check your local city website to find out if these services exist near you.
2. In addition, reduce your food waste by shopping with a grocery list. Intentionally planning your meals significantly reduces how much spoilage occurs in your fridge. Eat leftovers for lunch the next day or cook a little bit less to see if you can avoid having leftovers at all.
3. Create new meals with leftovers so it doesn't feel like you are having a repeat meal. For example, turn leftover veggies into a soup, which you can puree to a creamy consistency, especially if you add a potato for extra smoothness. Serve this with toasted bread and olive oil. Leftover rice can be turned into fried rice with a little soy sauce and ginger. Get creative and you'll be surprised at how much less food your family discards.
4. Lastly, many of us believe that we can eat way more than we actually can, and we can partially blame this on our plate sizes, which have grown significantly since the 1960s. In fact, research shows that the size of a typical dinner plate has increased by almost 36 percent since 1960.[11] This may not sound like much, but when each meal of each day is on a larger plate, we lose sight of what an appropriate portion size actually looks like. By using smaller plates, we take much less food and have a better chance of finishing what is on our plates. A study published in the *American Journal of Preventive Medicine* found that when people were given larger bowls, they served themselves larger portions.[12]

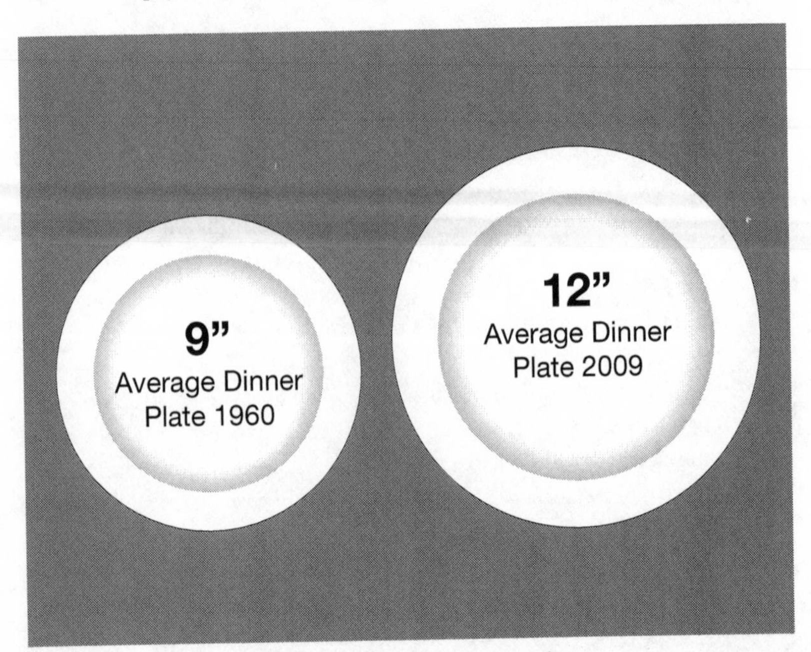

Figure 9.2. Change in the size of dinner plates over time.

Home Cleaning Supplies

The typical American family of four paid an average of $837 for housekeeping supplies in 2020. Thanks to extensive marketing and advertising dollars by the housekeeping industry, we have been led to believe that we cannot live without their products.[13] (Not to mention the whopping 2.5 billion plastic cleaning bottles that end up in landfills yearly.)[14] Follow these next tips to save your family from harsh and unnecessary chemicals while saving our planet from the billions of plastic containers discarded.

1. We can save money and plastic waste simply by switching to natural replacements like vinegar and water in a reusable spray bottle or by creating a cleaning paste with baking soda and lemon juice or vinegar. Get creative by adding lavender or other essential oils while experimenting with other natural options that are just as strong as the chemical counterparts. If you don't believe this, just give it a try once and compare. My favorite is

the one-to-one mix of white vinegar and water, which I learned about while watching a waiter in a restaurant remove water spots from wineglasses. I had to know what this magic potion was, which was safe to be used on our drinking glasses, and he told me that it's the best glass cleaner in the world. I use this handy solution to clean not only all glass tabletops and windows but also the bathtub, toilet, sink, floors, and all countertops.

2. Another cleaning go-to is baking soda and hydrogen peroxide. Mix one-quarter cup baking soda in a small glass bowl and add hydrogen peroxide to create a thick paste. This mixture will scrub grime away from any stubborn surface, including in greasy pans or the oven. For example (and one of my favorites), I proved to my mother-in-law that a pan I had was not trash by boiling some baking soda and water and giving the pan a scrub. Before long, it was as good as new.
3. You can even unclog your drain with a cup of baking soda followed by one cup of heated vinegar.

Self-Care and Hygiene

1. Try new toothpaste products, such as small pellets that you can pop in your mouth and start brushing, instead of buying plastic tubes. The market size of the global toothpaste industry was $35.5 billion in 2022 and is expected to grow at a compound annual growth rate of 6.2 percent into 2030.[15] "In America alone, consumers use more than 400 million tubes of toothpaste annually. Most end up in landfills because the multi-material packaging isn't generally accepted at recycling centers."[16]
2. Try refillable dental floss! Some great brands out there are using small, refillable glass containers that are even better than the plastic kind. And, believe me, it adds up. "Over three million miles' worth of plastic dental floss is used and thrown out in North America every year. That's the equivalent of traveling to the moon and back over six times."[17] This total doesn't even include the newer plastic flossing picks, which add up to 4.7 billion pieces of disposable single-use plastic tossed out every year![18]

3. Switch out plastic toothbrushes for bamboo ones. According to *National Geographic*, we discard more than one billion plastic toothbrushes every year just in the United States.[19] Moreover, research shows that plastic can take up to a thousand years to break down. On the other hand, natural materials like bamboo may take around six months to break down.[20]
4. Switch to reusable razors instead of contributing to the two billion razors thrown away each year.[21] While reusable razors are pricier upfront, you will save you hundreds in the long run and keep plastic out of the environment. Compared to conventional plastic razors, which are rarely recycled, reusable safety razors are more sustainable because you can replace just the blade.
5. Ladies, consider exploring sustainable menstrual products. Many new options are available, from washable period underwear and washable pads to menstrual cups. Even tampons made from natural fibers and without applicators are a better choice than plastic-wrapped tampons with plastic applicators, which are found during beach cleanups worldwide. The feminine product industry is anything but sustainable. In the United States alone, approximately twelve billion pads and seven billion tampons are discarded each year. Most American women menstruate for about forty years in total, using somewhere between five thousand and fifteen thousand pads and tampons, the vast majority of which wind up in landfills as plastic waste.[22]
6. Skip baths and keep your showers to five minutes. According to the EPA, a full bathtub requires about seventy gallons of water, while taking a five-minute shower uses about ten gallons of water. Expect to use about two gallons of water each minute.[23] It adds up quickly. "A shower lasting about 5 minutes emits between 90 and 200 grams of CO_2 if using a gas boiler, and as much as 500 grams if an electric boiler is used."[24]
7. Make sure the faucet is off while brushing your teeth. This is something that we imagine everyone already knows, but that is not the case! I have been left dumbfounded on many trips with friends as my roommates left the water on the entire time they brushed their teeth, even walking away to grab something from their suitcase. According to the EPA, "turning off the tap while

brushing your teeth can save 8 gallons of water per day."[25] That means a family of four can save thirty-two gallons of water each day!

8. Try the bar form of soaps, shampoos, and conditioners. As mentioned in earlier chapters, since industries study our buying habits in order to forecast changes and trends, we can effect positive change by informing them of the direction we want them to take when we try out these new forms of products. We hold the power. We must use this power and campaigns like #BarsOverBottles to kick-start these opportunities for making change.

Fashion

Most people aren't aware of the impact the $2.5 trillion fashion industry has on our planet. Below are some eye-opening facts and figures.[26]

- Each second, the equivalent of a garbage truck's worth of clothes are discarded and burned.
- Around 60 percent of the fashion industry's materials are plastic based.
- Microfibers released from the fashion industry are equal to the pollution of fifty billion plastic bottles, with over five hundred tons released each year.
- The fashion industry's carbon emissions, about 8–10 percent of humanity's carbon emissions, are more than those from all international flights and maritime shipping combined.
- Each year the fashion industry uses ninety-three billion cubic meters of water, contributing to water scarcity.
- The fashion industry is responsible for around 20 percent of industrial wastewater pollution.

1. Avoid fast fashion. A study showed that "while people bought 60 percent more garments in 2014 than in 2000, they only kept the clothes for half as long."[27] Huge corporations are coming out with clothing that is inexpensive and marketed as practically "single use." We must normalize repeating outfits and place value on buying fewer items that can pass the test of time

and be passed down to our kids. (I enjoy visiting my family in Europe and watching how proudly families keep clothing and other heirlooms for future generations. The children are always dressed so beautifully because great-quality clothing is built to last.)

2. Go thrifting! This is an easy way to participate in a circular economy by which the life cycle of an item is extended through trading, recycling, sharing, repairing, and other methods, while you get to fill your wardrobe with something fun and new. Thrifting doesn't have to be for low-end items only. Incredible vintage stores that have high-end brands are everywhere these days.
3. Always donate unwanted or unneeded clothing to homeless shelters. They need them the most, and they can really use winter items in particular. Coats and jackets are always welcome to handle harsh weather.
4. Fix it instead of tossing it! You don't need to be perfect at all. In fact, there is a whole movement based on the centuries-old Japanese tradition of Sashiko, a form of needlework in which stitched patterns and designs are added to older articles of clothing to strengthen the fabric. Individuals concerned about the environmental impact of new clothing are driving this "visible mending" movement.[28] By making small repairs to the clothing you already own, you can wear it longer and show off a little bit of character—newer doesn't always mean always better.

Your Garden

1. Opt for native flowers that require less water. Native plants have adapted to local environmental conditions and require far less water, saving time, money, and perhaps the most valuable natural resource, water. Native plants benefit many species of wildlife and provide birds with vital habitat.[29]
2. Grow your own herbs. Herbs typically come packaged in single-use-plastic clamshells, which for the most part are not recycled, generating a ton of unnecessary plastic.[30] With a bit of soil, sun, and water you can grow your own herbs nearly anywhere.

3. Save water by placing a bucket in the shower as you wait for the running water to heat up. You can then use this to water your garden and plants. Remember, you can get two gallons of water for every minute your shower is turned on.

Food: Our Food Matters the Most!

1. Buy local! Most cities have farmers markets with seasonal produce. Did you know that the average carrot bought at a supermarket traveled over two thousand miles to get there? Compare that with your local farmers market, which is usually no more than fifty to a hundred miles away.

    A study called "Food, Fuel, and Freeways" released by the Leopold Center for Sustainable Agriculture in Iowa compiled data from the U.S. Department of Agriculture to find out how far produce traveled to a Chicago "terminal market."[31] Table 9.1 contains a few staggering distances our produce travels to arrive at our supermarkets versus farmers markets. (Keep in mind that in the United States, the average freight truck gives off 161.8 grams of carbon dioxide per ton-mile.[32])

Table 9.1. Average Distance (in miles) Produce Travels from Farm to Supermarket versus Farmers Market

Produce	*Supermarket*	*Farmers Market*
Grapes	2,143	134
Lettuce	2,055	102
Peaches	1,674	173
Apples	1,555	77
Tomatoes	1,369	117
Greens	889	99
Winter Squash	781	98
Beans	766	101

2. Skip animal products by starting with at least eight plant-based meals per week, which will decrease your overall animal product consumption by 40 percent. Download the Habits of Waste app to help you easily adopt a partially plant-based diet.

3. Take a zero-waste approach when cooking. For example, carrot tops make great pesto sauce, and asparagus ends can be turned into a delicious soup! Google the item you have on hand and search recipes to create interesting ways to cook it.
4. BYO lunch to work so that you don't get so many plastic containers when ordering in. I purposely cook a little bit of extra dinner so I can take the leftovers to work.
5. Avoid *all* products with palm oil. This one is essential because palm oil has been and continues to be a major driver of deforestation of some of the world's most biodiverse forests. Palm oil is the reason for destroying the habitats of already endangered species like the orangutan, pygmy elephant, and Sumatran rhinoceros. "Every minute, an area the size of 300 football fields of the world's most dense, species rich forest is destroyed to create palm oil plantations. Palm oil is used in 1 of every 2 packaged food products found in supermarkets worldwide," so read the ingredients closely when shopping.[33]

Extra Credit

Join a "buy nothing, sell nothing" group! Helping out a neighbor has never been easier. These community groups exist on social media platforms like Facebook as a global conglomeration based on the Buy Nothing Project founded in 2013 in Bainbridge Island, Washington, to encourage giving (or recycling) of consumer goods and services over conventional commerce.

Remember, each of us is one person in a sea of almost eight billion people. If you live in the United States, your carbon dioxide emissions are among the highest in the world.

The bottom line is that we cannot wait for laws and policy to change if we don't speak up. We must first become aware of the changes we want made by trying to address the "habits of waste" in our lives. In addition, we can flex our buying power by showing our priorities through our dollars. How we spend our money defines the future because companies are spending billions to understand what we consumers want so they can react accordingly.

Slowly but surely, with a combination of our new greener habits and our demands for greener policies, we will create the grassroots uprising needed to force legislators and corporations to align their businesses with our values.

Finally, please share tips to kick those habits of waste with your friends and family, at your workplace, and on your social media networks. You are not alone in this, and by forming groups you can create a stronger force forward. This is about reaching critical mass—the minimum needed to create a reaction—and that takes time, patience, and diligence.

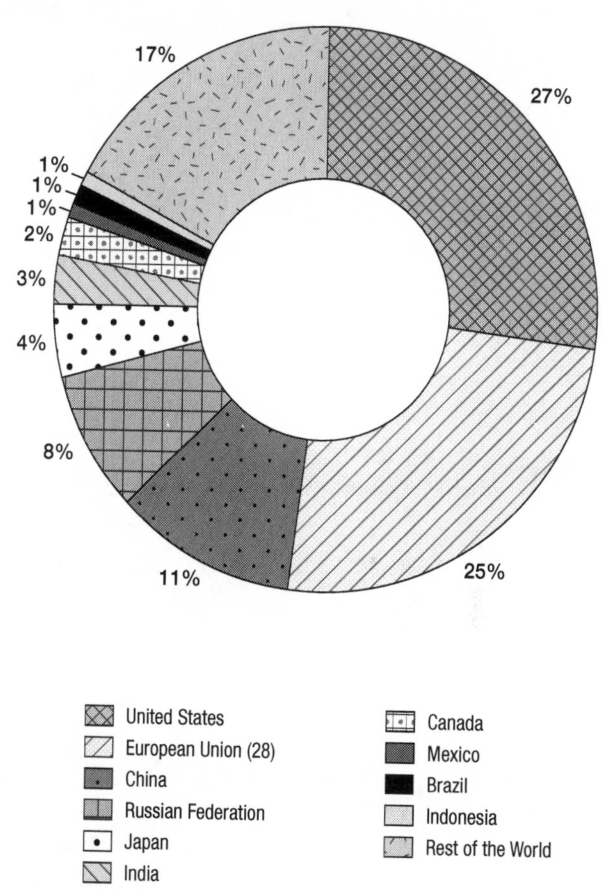

Figure 9.3. Cumulative carbon dioxide emissions from 1750 to 2019. *Source*: Our World in Data.

10

EPILOGUE

This chapter is a more personal one. I shared some of the details of my life with you because I wanted you to know that I am not very different from everyone else. In fact, I struggled a lot socially, financially, and emotionally. I felt "less than" for most of my life, until I found my inner power and realized that you don't have be an MIT or Harvard graduate to be considered smart or capable. In fact, these are hierarchical constructs that we have assigned to ourselves that only hold us back from being the best versions of ourselves.

I hope this book has inspired you so far. If you have any doubt in your mind that you can do similar things, then this chapter will convince you that you are able, capable, and in the position to be an imperfect environmentalist who can also be a massive changemaker.

The truth is, if I could do it, then anyone can. All you need is a healthy dose of persistence.

MY BACKSTORY

My story begins with my journey out of Tehran, Iran, during the 1979 Islamic Revolution. I don't remember much, but this event changed my life forever. You see, my parents weren't your typical Iranians. My father went to university in Florence, Italy, and became an architect. He returned to his homeland after graduation and was among the coolest bachelors in Tehran. My mother went to boarding school in Kent, England, and then spent a few years in Paris. She was a bombshell. They lived in an era where Iran was a place people dreamed of visiting. The Shah had created a little piece of history where Iran resembled Europe,

Figure 10.1. In nature. Photograph by Sheila Morovati.

and he did whatever he could to create a more modern regime with much Western political influence. From my family's perspective, it was a wonderful era to be in Iran. Ultimately, Iran was a country of the Middle East and had traditions and religious values that eventually overthrew the Shah and all his progressive ways. Iran became the Islamic Republic of Iran and reverted back to the days where women had virtually no rights at all. We are devastated by the ultimate fate of our beautiful country.

I was lucky that my parents were so progressive and had wanted me to live a better life. They didn't want me to end up as a wife of someone without my own career path. This was the driving force that made them hurriedly leave when it was announced that Ruhollah Khomeini would be the supreme leader of the country. They knew that my freedom depended on their escape from Iran, so they left with just a suitcase of clothes, a few thousand dollars, and me, their nine-month-old baby girl. We moved around a lot. I don't remember anything until we arrived in Fort Lee, New Jersey, which is a small town across the river from Manhattan. The Hudson River is what separated us from the Big Apple. Like many immigrants, my family would simply cross the George Washington Bridge daily to go to work in the city. My dad had to abandon his true passion, a career as an architect, because neither his Italian nor his Iranian architecture license was accepted in the United States. He had the choice to go back to school here in the United States or find a job immediately so that he could feed his family. There wasn't much of a choice, so he gave up his art and began working in order for us to have a nice apartment and a comfortable life. He knew how to buy and sell rugs and antiques because that is how he and his brother made a living while going to school in Italy. Later on, my dad shifted gears and created a men's fashion line, but until then it was always antiques and rugs.

Many Sundays we would journey to New Hampshire since that was where many of the auctions were held. The musty smell in these auction houses is something I will never forget. I witnessed people feverishly outbidding one another for old and dingy-looking furniture and rugs. It was here that I learned that shiny and new didn't always mean better. When my mom joined my dad, it was usually because she wanted to buy something for our little apartment. She had virtually no decorating budget, but we still needed a place to sit and eat and so forth. The first piece I remember them buying was an antique dining set. This was

Figure 10.2. In nature. Photograph by Sheila Morovati.

a pretty beat-up table, but it had good bones. We somehow took it apart and brought it home in my dad's 1970s brown Volvo station wagon. As soon as we entered the parking lot of our apartment building, my dad would let me sit in his lap and drive the car to our parking spot. This was one of the best parts of those trips to New Hampshire. Then I watched my dad work for days to refurbish this table and the chairs. Instead of spending thousands of dollars to professionally reupholster the chairs, he took stacks of my little sister's cloth diapers and flattened about six or eight of them to replace the worn-out cushions for each chair. Next, he re-covered the chairs with new fabric that my mom had found, and voilà, we had a pretty nice dining room table. I didn't know it then, but right there I witnessed my dad breathe new life into something that had looked so old and decrepit.

Being an energetic kid in a two-bedroom apartment on the seventeenth floor of a high-rise was tough. I wanted to go outside and play all the time, but there was no way to go outside. Luckily there was a little park across the street that had no jungle gym, swings, or anything else, but it had beautiful trees and a couple statues of canons made of bronze. These were fun to climb on, and my dad had the patience to let me spend time climbing those often. During our many outings he would take me for a walk in this park and teach me how to study the type of tree we were looking at based on its trunk. Cherry trees were the easiest to spot because they had small, horizontal lines all over its smooth bark. This was our activity at every park visit, which was usually on Sundays, my dad's only day off work. We would walk around this little park and find the cherry trees. My sister and I having time outside was a priority for him, and I recall that he would regularly tell my mom that our rowdiness inside the apartment was because we had pent-up energy and needed to go outside to release it. My dad was and still is a very progressive person who loves nature deeply.

For our sake he was determined to buy a house with a yard. One day, we went for a drive in a suburban area twenty minutes away from Fort Lee. This was a town called Closter and it was filled with trees and the most beautiful spacious lawns. We saw a home with a "for sale by owner" sign on the front lawn. It was only 8:00 a.m., but my dad knocked on the door and asked if we could see the home. The owners were an elderly couple, who were definitely surprised to see this young family of four looking to potentially buy their home. Even though they were in the middle of their breakfast, they welcomed us and gave us a tour. I saw my mom's jaw drop when my dad announced to the owners that he wanted to buy this house. Within thirty minutes he gave them a deposit and the ball started rolling on our first home in the United States. It was all fueled by his wish for us to go outside whenever we wanted.

This house was so special. The house itself was small but comfortable. It was a ranch-style home with three bedrooms and two bathrooms, a formal living room, dining room, and den, plus a kitchen and breakfast nook. There were windows everywhere, and the light that came through this house was stunning. The outside was exceptional, especially to my eyes. This home sat on one acre of land and had a creek running along the side of it. This creek became my best friend, and it is where I spent most of my childhood. This creek supported me through some of the toughest times of my life, and I realized there that nature has the power to heal like no other.

At this point, I was about eight years old and had already switched several schools. My parents thought that it would be a great idea for me to skip preschool and kindergarten and enter first grade as a five-year-old. Their reasoning was simply because I was tall and mature for my age, and because their friend's daughter would be in the same grade. In addition, this is a typical "tiger parent" thing to do for Iranians, where they push their children ahead prematurely just for bragging rights that their child is gifted and has skipped ahead. Sadly for my parents, their decision to put me ahead was not well thought out because I was virtually failing school.

School lunch breaks were the toughest time for me growing up. My mom usually packed me Persian food for lunch, and all the children would point and laugh that my lunch looked so funny and smelled so different. I asked my mom for peanut butter and jelly in an effort to fit in, even though I found it to be such a difficult consistency to chew and swallow. Assimilation was my goal, so I ate those sticky PB&J sandwiches as one step toward making friends. Of course, my mom had to get the healthiest whole wheat brown bread with some crunchy peanut butter that had oil sitting at the top, which you had to stir in. The kids at school got white bread with some smooth peanut butter and grape jelly,

Figure 10.3. Together with my family in nature. Photograph by Sheila Morovati.

Figure 10.4. We can enjoy many different activities in nature! For example, have your next meal outdoors. Photograph by Sheila Morovati.

which looked much better than my brown, dry sandwich with clumps of peanuts to chew through on top of everything else.

Next I went to Hillside School, a sweet public school where I was happier at first but soon realized that I was never going to make it with the kids being a whole year older than me. The kids were definitely cruel because I was a year younger, which meant I was the runt of the group. Eventually the school insisted that I transition back to third grade with kids my own age. I was so embarrassed by being "left back." Luckily a woman by the name of Sheila Hoffman met with me and was one of the first adults to be on my side. She made me feel proud to repeat third grade as a warrior of fourth grade and to share how it was going to be for everyone the next year. As I went back to third grade midyear, I shared all the details with my new classmates. I looked like a hero from the future.

School was still miserable for me, though. I recall going to the bathroom *a lot* because I was just so bored in class. I joined the chess team, which was interesting, but in general school was a major drag both socially and academically. I just thought differently from the rest of the kids. For example, one of our assignments in English class was to create a report on magazines and how they operate as a publication. I decided that it would be more interesting if I created my own magazine and wrote articles and placed images and "ads" on every page to really show what it would take to publish one issue of a magazine from start to finish. After spending hours on this assignment, writing sample articles, gluing images, creating ads, and binding it, I turned it in with so much pride and excitement. But I ended up earning a "D" on it. I was in total shock. When I spoke to the teacher about it, she said I hadn't followed the directions.

I would find solace in the creek by my house, spending hours climbing down the bank and jumping across to the other side by carefully placing my feet strategically on the rocks that poked out of the water. There was something about this creek that made everything feel better and more manageable.

By the time I reached fifth grade my family decided that this would be my last year at Hillside School. It was time to leave New Jersey because my mom's brothers in San Jose, California, had started a very successful construction company and could use my dad's help, with his architectural background, which also meant he could go back to doing what he loved. It didn't take much to convince my parents to move

Figure 10.5. In nature. Photograph by Sheila Morovati.

to California. I was thrilled because I would get a fresh start in a new school, plus I'd be surrounded by family that I loved so much.

California was better overall. I was happier and found a few nice friends. It was a time when my dad had the chance to resurrect his love of architecture. But within two years, in 1989, the recession hit and my family had to move once again.

This time we ended up in Beverly Hills, California. My parents opened a very elegant men's clothing store and rented a sweet little house with a picket fence on a very busy street. The home had virtually no yard, but we did have one lemon tree and a couple rosebushes.

I could feel the disconnect from nature really start to take a toll on me. I was living in a very urban area and went to school at Beverly Hills High and then to college at UCLA. It was city life for sure and didn't have the solace of my old creek anywhere. At the time, I could not put my finger on what was missing, but looking back on my life I see clearly that I needed more time immersed in nature.

High school continued to feel like torture. I spent most of my days watching the clock tick by each minute, wishing that 3:00 p.m. would come sooner just so I could escape. When I was a senior in high school, my dad and I had a heart-to-heart about what major I should pick in college, and he explained that I should follow my interests without thinking of what I could do with it next. I chose sociology because I thought studying the psychology of the masses was fascinating. It was the right choice because, luckily, I fell in love with it, and while I sat in those classes I didn't look at the clock once. In fact, I was hanging onto the professors' lectures and taking notes vigorously. It was at UCLA that I finally saw the potential I had inside. The recommendation was that undergraduate students take three classes per quarter. But I wanted more, so I took four classes each quarter and attended summer school. I graduated magna cum laude in three years. Looking back, I wish I hadn't rushed so much.

After graduating I went to work for an online advertising firm and made hardly any money. It was the dot-com boom at the time, and I learned so much. All the catchy campaigns I created for Habits of Waste come from a marketing and advertising mindset because I know what it takes to infiltrate society with a product or an idea.

I got married in 2004 and had my daughter in 2007. My husband and I started out in an apartment, but soon after Sofia was born we decided to move to Pacific Palisades, which is a small beach community sandwiched between Malibu and Santa Monica. We wanted to move into a home with a yard so that she could have more space to be outside and surrounded by nature. History was repeating itself.

Our new home was situated high in the mountains with an expansive view of the ocean. The serenity of this area is incredible. Almost as soon as we moved in I slowly started reconnecting with my first love, nature. I began to revive my sense of solace by hiking in the mountains or taking long walks on the beach. Part of my mission is to tell people to go outside and connect with nature because you cannot believe how supportive nature is to all of us. I still hike with my best friend at least once a week on the Temescal Canyon trailhead. I call it the transformation loop because as you enter the hike toward the right side you are holding on to many thoughts, worries, and stress, but by the time you

Figure 10.6. In Los Angeles, where I live. Photograph by Sheila Morovati.

complete the loop and come out the other side you feel that your mind has been cleared and that your conscience is lighter. It's as though the trees, the views, and the sounds pull all the negativity out of your body, replacing it with tranquility.

I wish more people could see how close we all are to this planet and how interconnected all of nature is to all of us humans. Nature is an extension of our bodies. I sincerely believe that if more people felt this connection, they would be more conscious of their actions and the impact we make on the planet. Indigenous peoples have had this connection to the Earth since the beginning, and they see themselves as a part of the same system as the environment in which we cohabitate. There is no sense of entitlement; rather, natural resources are considered valuable, shared property, and they are respected as such. I think we can all learn a few great lessons from the Indigenous peoples.

One of my goals in writing this book was to give every reader the opportunity to recognize the power they hold within them. I want everyone to know that they too can create change in their home, school, community, city, or even state. By now, hopefully, you are able to see the world around you differently—you have a role to play, an imperative role, to be a part of the solution to protect our planet.

I still wake up each and every morning with the mission to protect our planet in any and every way I can think of. Not everyone will have a path like mine, but one thing is for sure: Nature has never let me down, and I will not ever let nature down. Join me in protecting our planet by eliminating one "habit of waste" at a time.

NOTES

CHAPTER 1

1. Hellmann, Kai-Uwe, and Marius K. Luedicke, "The Throwaway Society: A Look in the Back Mirror," *Journal of Consumer Policy* 41, no. 1 (2018): 83–87, https://doi.org/10.1007/s10603-018-9371-6.

2. U.S. Environmental Protection Agency, "National Overview: Facts and Figures on Materials, Wastes and Recycling," Overviews and Factsheets, October 2, 2017, https://www.epa.gov/facts-and-figures-about-materials-waste-and-recycling/national-overview-facts-and-figures-materials.

3. Hellmann and Luedicke, "The Throwaway Society."

4. Cheyenne Mountain Zoo, "What Happens to All Those Unwanted, Broken or Much Loved Crayons?" August 30, 2019, https://www.cmzoo.org/news/archive/what-happens-to-all-those-unwanted-broken-or-much-loved-crayons/.

5. Litvinov, Amanda, "Out-of-Pocket Spending on School Supplies Adds to Strain on Educators," National Education Association, October 14, 2022, https://www.nea.org/advocating-for-change/new-from-nea/out-pocket-spending-school-supplies-adds-strain-educators.

6. Tamer, Mary, "On the Chopping Block, Again," *Harvard Ed. Magazine*, June 8, 2009, https://www.gse.harvard.edu/news/ed/09/06/chopping-block-again.

7. Crayon Collection, "Crayon Recycling Program," accessed May 22, 2023, https://crayoncollection.org/programs/crayon-recycling/.

8. Crayola, "How Many Crayola Products Do You Make per Year?" Accessed May 19, 2023, https://www.crayola.com/faq/your-business/how-many-crayola-products-do-you-make-per-year/.

9. U.S. Department of Education, Title I, Part A Program, November 7, 2018. https://www2.ed.gov/programs/titleiparta/index.html.

10. Vargas, Annette M., "Arts Education Funding," *Journal of Women in Educational Leadership* (2017), https://doi.org/10.13014/K21Z42K0.

11. Chen, Grace, "How the Arts Benefit Your Children Academically and Behaviorally," Public School Review, February 13, 2023, https://www.publicschoolreview.com/blog/how-the-arts-benefit-your-children-academically-and-behaviorally.

12. Chen, "How the Arts Benefit Your Children Academically and Behaviorally."

13. Kim, Jinho, and Kerem Morgül, "Long-Term Consequences of Youth Volunteering: Voluntary Versus Involuntary Service," *Social Science Research* 67 (September 2017): 160–75, https://doi.org/10.1016/j.ssresearch.2017.05.002.

CHAPTER 2

1. Gao, Angela L., and Yongshan Wan, "Life Cycle Assessment of Environmental Impact of Disposable Drinking Straws: A Trade-off Analysis with Marine Litter in the United States," *Science of the Total Environment* 817 (2022): 153016, https://doi.org/10.1016/j.scitotenv.2022.153016.

2. Gao and Wan, "Life Cycle Assessment of Environmental Impact of Disposable Drinking Straws."

3. Lonely Whale, "#StopSucking | Lonely Whale | For A Strawless Ocean," YouTube, 2017. https://www.youtube.com/watch?v=Q91-23B8yCg.

4. Sea Turtle Biologist, "Sea Turtle with Straw up Its Nostril—'NO' TO SINGLE-USE PLASTIC," YouTube, 2015, https://www.youtube.com/watch?v=4wH878t78bw.

5. Booker, Linda, director, *Straws* (By the Brook Productions, 2017).

6. Rujnić Havstad, Maja, Ljerka Juroš, Zvonimir Katančić, and Ana Pilipović, "Influence of Home Composting on Tensile Properties of Commercial Biodegradable Plastic Films," *Polymers* 13, no. 16 (2021): 2785, https://doi.org/10.3390/polym13162785.

7. CalRecycle, "Compost and Mulch Producers," accessed May 26, 2023, https://calrecycle.ca.gov/organics/processors/.

8. Carnegie, Dale, *How to Win Friends and Influence People*, revised edition (New York: Pocket Books, 1982).

9. Dykstra, Alan, "The Rhetoric of 'Whataboutism' in American Journalism and Political Identity," *Res Rhetorica* 7, no. 2 (2020), https://doi.org/10.29107/rr2020.2.1.

CHAPTER 3

1. Groeger, Lena V., "Set It and Forget It: How Default Settings Rule the World," ProPublica, July 27, 2016, https://www.propublica.org/article/set-it-and-forget-it-how-default-settings-rule-the-world.

2. AppsRhino, "Food Delivery App | User Statistics 2022," March 29, 2023, https://www.appsrhino.com/blogs/food-delivery-app-user-statistics-2022.

3. Stockton Recycles, "Plastic Utensils Go in the Trash," March 5, 2020, https://stocktonrecycles.com/plastic-utensils-go-in-the-trash/.

4. U.S. EPA, "Plastics: Material-Specific Data," Collections and Lists, September 12, 2017, https://www.epa.gov/facts-and-figures-about-materials-waste-and-recycling/plastics-material-specific-data.

5. Gautam, Anirudh Muralidharan, and Nídia Caetano, "Study, Design and Analysis of Sustainable Alternatives to Plastic Takeaway Cutlery and Crockery," *Energy Procedia* 136 (October 2017): 507–12, https://doi.org/10.1016/j.egypro.2017.10.273.

6. Curry, David, "Uber Eats Revenue and Usage Statistics (2023)," Business of Apps, May 2, 2023, https://www.businessofapps.com/data/uber-eats-statistics/.

7. SameSide, accessed May 30, 2023, https://onsameside.com/.

8. Goodful, "Everybody's Dirty Little Secret," Facebook, November 20, 2020, https://www.facebook.com/watch/?v=156800712810150.

9. Birnbaum, Michael, "Plastic Forks Become Rarer as England Joins Global Effort to Ban Them," *Washington Post*, January 10, 2023, https://www.washingtonpost.com/climate-solutions/2023/01/10/england-single-use-cutlery-ban/.

10. AppsRhino, "Food Delivery App | User Statistics 2022."

CHAPTER 4

1. Bouyé, Mathilde, and David Waskow, "Climate Action Isn't Reaching the Most Vulnerable—But It Could," November 11, 2021, https://www.wri.org/insights/how-climate-action-can-help-vulnerable-populations.

2. Mahoney, Adam, "In America's Cities, Inequality is Engrained in the Trees," Grist, May 5, 2021, https://grist.org/cities/tree-cover-race-class-segregation/. See also McDonald, Robert I., Tanushree Biswas, Cedilla Sachar, Ian Housman, Timothy M. Boucher, Deborah Balk, David Nowak, Erica Spotswood, Charlotte K. Stanley, and Stefan Leyk, "The Tree Cover and Temperature Disparity in U.S. Urbanized Areas: Quantifying the Association with Income across 5,723 Communities," *PloS One* 16, no. 4 (2021): e0249715, https://doi.org/10.1371/journal.pone.0249715.

3. Rudd, Lauren F., Shorna Allred, Julius G. Bright Ross, Darragh Hare, Merlyn Nomusa Nkomo, Kartik Shanker, Tanesha Allen, et al. "Overcoming Racism in the Twin Spheres of Conservation Science and Practice," *Proceedings of the Royal Society B: Biological Sciences* 288, no. 1962 (2021): 20211871, https://doi.org/10.1098/rspb.2021.1871.

4. U.S. EPA, "EPA Report Shows Disproportionate Impacts of Climate Change on Socially Vulnerable Populations in the United States," September 2, 2021, https://www.epa.gov/newsreleases/epa-report-shows-disproportionate-impacts-climate-change-socially-vulnerable.

5. Moran, Alyssa J., Aviva Musicus, Mary T. Gorski Findling, Ian F. Brissette, Ann A. Lowenfels, S. V. Subramanian, and Christina A. Roberto, "Increases in Sugary Drink Marketing During Supplemental Nutrition Assistance Program Benefit Issuance in New York," *American Journal of Preventive Medicine* 55, no. 1 (July 2018): 55–62, https://doi.org/10.1016/j.amepre.2018.03.012.

6. Centers for Disease Control and Prevention, "Increase Access to Drinking Water in Schools," April 16, 2021, https://www.cdc.gov/healthyschools/features/water_access.htm.

7. Moreno, Gala D., Laura A. Schmidt, Lorrene D. Ritchie, Charles E. McCulloch, Michael D. Cabana, Claire D. Brindis, Lawrence W. Green, Emily A. Altman, and Anisha I. Patel, "A Cluster-Randomized Controlled Trial of an Elementary School Drinking Water Access and Promotion Intervention: Rationale, Study Design, and Protocol," *Contemporary Clinical Trials* 101 (February 2021): 106255, https://doi.org/10.1016/j.cct.2020.106255.

8. Bloom, Richard, Marc Levine, and Robert Rivas, "School Facilities: Drinking Water: Water Bottle Filling Stations," Pub. L. No. AB 2638, California Legislative Information (2022), https://leginfo.legislature.ca.gov/faces/billTextClient.xhtml?bill_id=202120220AB2638.

9. Los Angeles Unified School District, "Los Angeles Unified Approves Infrastructure Upgrades and Sustainable Green Outdoor Learning Spaces," March 27, 2023. https://www.lausd.org/site/default.aspx?PageType=3&DomainID=4&ModuleInstanceID=4466&ViewID=6446EE88-D30C-497E-9316-3F8874B3E108&RenderLoc=0&FlexDataID=131211&PageID=1.

10. Langin, Katie, "Millions of Americans Drink Potentially Unsafe Tap Water. How Does Your County Stack Up?" *Science*, February 12, 2018. https://www.science.org/content/article/millions-americans-drink-potentially-unsafe-tap-water-how-does-your-county-stack.

11. Schuman, Rebecca, "This Viral Formula Ad Absolves You for Using Formula," *Slate*, June 4, 2015, https://slate.com/human-interest/2015/06/a-century-of-formula-advertising-its-always-gone-straight-for-the-new-mother-jugular.html.

12. *New York Times*, "F.D.A Is Sued on Drug to Dry Mothers' Milk," August 17, 1994, sec. U.S., https://www.nytimes.com/1994/08/17/us/fda-is-sued-on-drug-to-dry-mothers-milk.html.

13. Foss, Katherine A., and Brian G. Southwell, "Infant Feeding and the Media: The Relationship between *Parents' Magazine* Content and Breastfeeding, 1972–2000," *International Breastfeeding Journal* 1, no. 10 (2006): 10, https://doi.org/10.1186/1746-4358-1-10.

14. Parker, Laura, "How the Plastic Bottle Went from Miracle Container to Hated Garbage," *National Geographic*, August 23, 2019. https://www.nationalgeographic.com/environment/article/plastic-bottles.

15. National Resources Defense Council, "The Truth About Tap," January 5, 2016, https://www.nrdc.org/stories/truth-about-tap.

16. Lazarus, David, "Column: You Do Know That, in Most Cases, Bottled Water Is Just Tap Water?" *Los Angeles Times*, September 28, 2021, https://www.latimes.com/business/story/2021-09-28/bottled-water-is-really-just-tap-water.

17. Majd, Sanaz, "Should You Drink Tap or Bottled Water?" *Scientific American*, October 21, 2015, https://www.scientificamerican.com/article/should-you-drink-tap-or-bottled-water/.

18. EARTHDAY.org, "Fact Sheet: Single-Use Plastics," March 29, 2022, https://www.earthday.org/fact-sheet-single-use-plastics/.

19. EARTHDAY.org, "Money in a Bottle," January 30, 2015, https://www.earthday.org/money-in-a-bottle/.

20. Los Angeles Department of Water and Power, "Plans to Install, Refurbish 200 Hydration Stations in Los Angeles Announced at 5th Annual Tap Water Day LA," June 13, 2019, https://www.ladwpnews.com/plans-to-install-refurbish-200-hydration-stations-in-los-angeles-announced-at-5th-annual-tap-water-day-la/.

CHAPTER 5

1. Igini, Martina, "The Environmental Impact of Online Shopping," Earth.Org, December 15, 2022. https://earth.org/online-shopping-and-its-environmental-impact/.

2. Goodchild, Anne, and Rishi Verma, "How Many Amazon Packages Get Delivered Each Year?" Supply Chain Transportation and Logistics Center at the University of Washington, October 17, 2022, https://depts.washington.edu/sctlctr/news-events/in-the-news/how-many-amazon-packages-get-delivered-each-year#.

3. Peters, Adele, "Can Online Retail Solve Its Packaging Problem?" Fast Company, April 20, 2018, https://www.fastcompany.com/40560641/can-online-retail-solve-its-packaging-problem.

4. Oceana, "Amazon's Plastic Problem Revealed," December 15, 2020, https://doi.org/10.5281/ZENODO.4341751.

5. Fortune, "United Parcel Service," accessed May 31, 2023, https://fortune.com/company/ups/.

6. Dean, Brian, "TikTok User Statistics (2023)," Backlinko, March 27, 2023, https://backlinko.com/tiktok-users.

7. Californians Against Waste, "AB 2026," accessed June 1, 2023, https://www.cawrecycles.org/ab-2026.

8. Oceana, "The Cost of Amazon's Plastic Denial on the World's Oceans," December 15, 2022, https://doi.org/10.5281/ZENODO.7525829.

CHAPTER 6

1. Howarth, Josh, "The Ultimate List of Beauty Industry Stats (2023)," Exploding Topics, March 23, 2023, https://explodingtopics.com/blog/beauty-industry-stats.

2. Ahmadi, Tala, "Replace Your Shampoo Bottle with a Shampoo Bar," The Boar, January 28, 2023, https://theboar.org/2023/01/replace-your-shampoo-bottle-with-a-shampoo-bar/#.

3. U.S. EPA, "Plastics: Material-Specific Data."

4. Demeneix, Barbara A., "How Fossil Fuel-Derived Pesticides and Plastics Harm Health, Biodiversity, and the Climate," *The Lancet Diabetes & Endocrinology* 8, no. 6 (June 2020): 462–64, https://doi.org/10.1016/S2213-8587(20)30116-9.

5. Campos, Paulina, "9 Beauty Ingredients That Are Banned in Europe (But Legal in the U.S.)," *Byrdie*, January 28, 2022, https://www.byrdie.com/banned-ingredients-europe.

6. Srinivasulu, M., M. Subhosh Chandra, Jaffer Mohiddin Gooty, and A. Madhavi, "Chapter 8—Personal Care Products—Fragrances, Cosmetics, and Sunscreens—in the Environment," in *Environmental Micropollutants*, edited by Muhammad Zaffar Hashmi, Shuhong Wang, and Zulkfil Ahmed, 131–49, Advances in Pollution Research (Amsterdam: Elsevier, 2022), https://doi.org/10.1016/B978-0-323-90555-8.00015-5.

7. National Oceanic and Atmospheric Administration (NOAA), "Sunscreen Chemicals and Marine Life," National Ocean Service, August 17, 2022. https://oceanservice.noaa.gov/news/sunscreen-corals.html.

8. Campos, "9 Beauty Ingredients That Are Banned in Europe."

CHAPTER 7

1. Center for Biological Diversity, "10 Facts About Single-Use Plastic Bags," accessed June 22, 2023, https://www.biologicaldiversity.org/programs/population_and_sustainability/sustainability/plastic_bag_facts.html.

2. Surfrider Foundation, "City of Los Angeles Plastic Bag Ban," June 6, 2013, https://www.surfrider.org/campaigns/City%20of%20Los%20Angeles%20Plastic%20Bag%20Ban.

3. Napurano, Clay, "California: Nation's First Statewide Plastic Bag Ban Cuts Waste," Frontier Group, August 27, 2021, https://frontiergroup.org/articles/california-nations-first-statewide-plastic-bag-ban-cuts-waste/.

4. Thompson, Don, "Think Those Bags Are Recyclable? Think Again," Associated Press, December 29, 2022, https://apnews.com/article/california-state-government-business-climate-and-environment-59dc96edb1b2901f829071a6f9153df1.

5. Plastic Oceans International, "Plastic Pollution Facts," accessed June 22, 2023, https://plasticoceans.org/the-facts/.

6. Narayanan, Rishya, "Does Plastic Come from Fossil Fuels? Yes—Here's What You Need to Know," Conservation Law Foundation, April 18, 2022, https://www.clf.org/blog/plastic-comes-from-fossil-fuels/. See also Gardiner, Beth, "The Plastics Pipeline: A Surge of New Production Is on the Way," *Yale Environment 360*, December 19, 2019, https://e360.yale.edu/features/the-plastics-pipeline-a-surge-of-new-production-is-on-the-way.

7. Barone, Emily, "U.S. Plastic Recycling Rates Are Even Worse Than We Thought," *Time*, May 19, 2022, https://time.com/6178386/plastic-recycling-rates-overestimated/.

8. Fearing, Franklin, "Influence of the Movies on Attitudes and Behavior," *The ANNALS of the American Academy of Political and Social Science* 254, no. 1 (November 1947): 70–79, https://doi.org/10.1177/000271624725400112.

9. World Wildlife Fund (WWF), "Will There Be More Plastic Than Fish in the Sea?" Accessed June 22, 2023, https://www.wwf.org.uk/myfootprint/challenges/will-there-be-more-plastic-fish-sea.

10. Statista, "Global: Traditional TV and Home Video Number of Users 2018–2027," June 1, 2023, https://www.statista.com/forecasts/1207931/tv-viewers-worldwide-number.

11. Jamieson, Patrick E., and Daniel Romer, "Portrayal of Tobacco Use in Prime-Time TV Dramas: Trends and Associations with Adult Cigarette Consumption—USA, 1955–2010," *Tobacco Control* 24, no. 3 (2015): 243–48, http://www.jstor.org/stable/24842472.

12. Karakartal, Demet, "Social Effects of Film and Television," *The Online Journal of New Horizons in Education* 11, no. 3 (2021), https://www.tojned.net/journals/tojned/articles/v11i03/v11i03-03.pdf.

13. Moraes, Marcela, John Gountas, Sandra Gountas, and Piyush Sharma, "Celebrity Influences on Consumer Decision Making: New Insights and Research Directions," *Journal of Marketing Management* 35, no. 13–14 (2019): 1159–92, https://doi.org/10.1080/0267257X.2019.1632373.

14. Newell, Jay, Charles T. Salmon, and Susan Chang, "The Hidden History of Product Placement," *Journal of Broadcasting & Electronic Media* 50, no. 4 (2006): 575–94, https://doi.org/10.1207/s15506878jobem5004_1.

15. Spielberg, Steven, director, *E.T. the Extra-Terrestrial* (Universal Pictures, 1982).

16. Statista, "Advertising Worldwide—Statistics and Facts," accessed June 22, 2023, https://www.statista.com/topics/990/global-advertising-market/.

17. Sony Pictures Entertainment, "Sony Pictures Entertainment to Eliminate Single-Use Plastic," August 3, 2020, https://www.sonypictures.com/corp/press_releases/2020/0803/sonypicturesentertainmenttoeliminatesingleuseplastic.

18. Pruner, Aaron, "Above-the-Line vs. Below-the-Line Jobs in Film," Backstage, May 10, 2022, https://www.backstage.com/magazine/article/above-the-line-vs-below-the-line-crew-differences-74969/. See also MasterClass, "Film Career Guide: Above the Line vs. Below the Line Jobs," June 24, 2021, https://www.masterclass.com/articles/film-career-guide-above-the-line-vs-below-the-line-jobs#.

19. Pener, Degen, "Universal Filmed Entertainment Group Launches GreenerLight Sustainability Program (Exclusive)," *The Hollywood Reporter*, March 9, 2023, https://www.hollywoodreporter.com/business/business-news/universal-filmed-entertainment-greenerlight-sustainability-program-1235346745/.

20. Fletcher, Angus, *Storythinking* (New York: Columbia University Press, 2023).

21. Gambino, Isabella, Francesco Bagordo, Tiziana Grassi, Alessandra Panico, and Antonella De Donno, "Occurrence of Microplastics in Tap and Bottled Water: Current Knowledge," *International Journal of Environmental Research and Public Health* 19, no. 9 (2022): 5283, https://doi.org/10.3390/ijerph19095283.

CHAPTER 8

1. Environmental Defense Fund, "Overfishing: The Most Serious Threat to Our Oceans," accessed June 21, 2023, https://www.edf.org/oceans/overfishing-most-serious-threat-our-oceans.

2. Mathieu, Edouard, and Hannah Ritchie, "What Share of People Say They Are Vegetarian, Vegan, or Flexitarian?" Our World in Data, May 13, 2022, https://ourworldindata.org/vegetarian-vegan.

3. Keoleian, Gregory, Martin Heller, and Diego Rose, "Implications of Future U.S. Diet Scenarios on Greenhouse Gas Emissions," Center for Sustainable Systems, University of Michigan, January 13, 2020, https://css.umich.edu/publications/research-publications/implications-future-us-diet-scenarios-greenhouse-gas-emissions.

4. Habits of Waste, "8meals," accessed June 21, 2023, https://habitsofwaste.org/campaigns/8meals/.

5. Holzer, Jillian, "Don't Put Vegetables in the Corner: Q&A with Behavioral Science Researcher Linda Bacon," World Resources Institute, June 12, 2017, https://www.wri.org/insights/dont-put-vegetables-corner-qa-behavioral-science-researcher-linda-bacon.

6. Veganuary, accessed August 3, 2023, https://veganuary.com/en-us/.

7. Davison, Courtney, "The Beginner's Guide to a Whole-Food, Plant-Based Diet," Forks Over Knives, July 28, 2023, https://www.forksoverknives.com/how-tos/plant-based-primer-beginners-guide-starting-plant-based-diet/.

8. Shieber, Jonathan, "The #8meals App from Habits of Waste Helps People Cut Back on Meaty Meals to Save the Planet," TechCrunch, April 25, 2021, https://techcrunch.com/2021/04/25/the-8meals-app-from-habits-of-waste-helps-people-cut-back-on-meaty-meals-to-save-the-planet/.

9. Mackintosh, Eliza, "The Amazon Is Burning Because the World Eats So Much Meat," CNN, August 23, 2019, https://www.cnn.com/2019/08/23/americas/brazil-beef-amazon-rainforest-fire-intl/index.html.

10. Cuffe, Sandra, "2022 Amazon Fires Tightly Tied to Recent Deforestation, New Data Show," Mongabay, November 22, 2022, https://news.mongabay.com/2022/11/2022-amazon-fires-tightly-tied-to-recent-deforestation-new-data-show/. See also Monitoring of the Andean Amazon Project, "MAAP #168: Amazon Fire Season 2022," November 3, 2022, https://maaproject.org/2022/amazon-fires-2022/.

11. World Wildlife Fund, "Unsustainable Cattle Ranching," accessed August 3, 2023, https://wwf.panda.org/discover/knowledge_hub/where_we_work/amazon/amazon_threats/unsustainable_cattle_ranching, citing data from Nepstad, Daniel C., Claudia M. Stickler, Britaldo Soares-Filho, and Frank Merry, "Interactions among Amazon Land Use, Forests and Climate: Prospects for a Near-Term Forest Tipping Point," *Philosophical Transactions of the Royal Society B: Biological Sciences* 363, no. 1498 (2008): 1737–46. https://doi.org/10.1098/rstb.2007.0036.

12. Nepstad et al., "Interactions among Amazon Land Use, Forests and Climate."

13. Generation Vegan, "Causes and Effects of Deforestation: How Can We Stop It?" GenV.org, February 9, 2021, https://genv.org/effects-of-deforestation/.

14. NASA Global Climate Change: Vital Signs of the Planet, "Which Is a Bigger Methane Source: Cow Belching or Cow Flatulence?" Accessed June 21, 2023, https://climate.nasa.gov/faq/33/which-is-a-bigger-methane-source-cow-belching-or-cow-flatulence.

15. Foster, Joanna, "Farmers, Scientists Seek Solutions to Global Warming Caused by Cows," Environmental Defense Fund, April 15, 2022, https://vitalsigns.edf.org/story/farmers-scientists-seek-solutions-global-warming-caused-cows.

16. Quinton, Amy, "Cows and Climate Change," University of California, Davis, June 27, 2019, https://www.ucdavis.edu/food/news/making-cattle-more-sustainable.

17. Columbia University Irving Medical Center, "What Is a Plant-Based Diet, and Is It Healthy?" April 15, 2022, https://www.cuimc.columbia.edu/news/what-plant-based-diet-and-it-healthy.

18. Grand View Research, "Plant-Based Meat Market Size, Share and Trends Analysis Report by Source (Soy, Pea, Wheat), by Product (Burgers, Sausages, Patties), by Type, by End-user, by Storage, by Region, and Segment Forecasts, 2023–2030," 2023, https://www.grandviewresearch.com/industry-analysis/plant-based-meat-market.

19. Poore, J., and T. Nemecek, "Reducing Food's Environmental Impacts through Producers and Consumers," *Science* 360, no. 6392 (June 1, 2018): 987–92, https://doi.org/10.1126/science.aaq0216.

CHAPTER 9

1. Mandel, Kyla, and Brad Plumer, "One Thing You Can Do: Smarter Laundry," *New York Times*, October 2, 2019, https://www.nytimes.com/2019/10/02/climate/nyt-climate-newsletter-laundry.html, citing data from Golden, Jay S., Vairavan Subramanian, Gustavo Marco Antonio Ugarte Irizarri, Philip White, and Frank Meier, "Energy and Carbon Impact from Residential Laundry in the United States," *Journal of Integrative Environmental Sciences* 7, no. 1 (2010): 53–73, https://doi.org/10.1080/19438150903541873.

2. U.S. Department of Energy, "LED Lighting," accessed August 3, 2023, https://www.energy.gov/energysaver/led-lighting.

3. U.S. EPA, "Tailpipe Greenhouse Gas Emissions from a Typical Passenger Vehicle," January 12, 2016, https://www.epa.gov/greenvehicles/tailpipe-greenhouse-gas-emissions-typical-passenger-vehicle.

4. Triantafillopoulos, Nick, and Alexander Koukoulas, "The Future of Single-Use Paper Coffee Cups: Current Progress and Outlook," *BioRes* 15, no. 3 (2020), accessed June 23, 2023, https://bioresources.cnr.ncsu.edu/resources/the-future-of-single-use-paper-coffee-cups-current-progress-and-outlook/

5. Newstalk, "Over 56 Billion Coffee Capsules to Go to Landfill This Year," November 26, 2018, https://www.newstalk.com/news/over-56-billion-coffee-capsules-to-go-to-landfill-this-year-492445.

6. Ramirez, Rachel, "Five Corporations Benefit from 25% of Bottled Water Sales," CNN, March 16, 2023, https://ix.cnn.io/charts/z6Wrz/2/.

7. Miller, Chaz, "Plastic Film," Waste360, October 1, 2008, https://www.waste360.com/Recycling_And_Processing/plastic_film_ldpe.

8. Bennett, Sophia, "How to Recycle Ziploc Bags," Recycle Nation, October 7, 2014, https://recyclenation.com/2014/10/recycle-ziploc-bags/.

9. Jarvie, Michelle, "Americans Waste Almost 50 Percent of Food Produced," Michigan State University Extension, November 21, 2014, https://www.canr.msu.edu/news/americans_waste_almost_50_percent_of_food_produced.

10. Jaglo, Kirsten, Shannon Kenny, and Jenny Stephenson, "From Farm to Kitchen: The Environmental Impacts of U.S. Food Waste," U.S. Environmental Protection Agency, November 2021, https://www.epa.gov/system/files/documents/2021-11/from-farm-to-kitchen-the-environmental-impacts-of-u.s.-food-waste_508-tagged.pdf.

11. Wansink, Brian, and Koert van Ittersum, "Portion Size Me: Downsizing Our Consumption Norms," *Journal of the American Dietetic Association* 107, no. 7 (2007), https://papers.ssrn.com/abstract=2474331.

12. Wansink, Brian, Koert van Ittersum, and James E. Painter, "Ice Cream Illusions: Bowls, Spoons, and Self-Served Portion Sizes," *American Journal of Preventive Medicine* 31, no. 3 (2006): 240–43, https://doi.org/10.1016/j.amepre.2006.04.003.

13. U.S. Bureau of Labor Statistics, "Consumer Expenditures in 2020," BLS Report 1096, December 2021, https://www.bls.gov/opub/reports/consumer-expenditures/2020/home.htm.

14. Tuman, Nina, "Why You Should Switch to Reusable Cleaning Bottles," Tru Earth, accessed June 24, 2023, https://www.tru.earth/Why-You-Should-Switch-to-Reusable-Cleaning-Bottles.

15. Goddiess, Samantha, "The 10 Largest Toothpaste Brands in the United States," Zippia, April 25, 2023, https://www.zippia.com/advice/largest-toothpaste-brands/.

16. Linnenkoper, Kirstin, "Colgate Leads Toothpaste Tube Recycling Innovation," Recycling International, June 28, 2019, https://recyclinginternational.com/plastics/colgate-leads-toothpaste-tube-recycling-innovation/26597/.

17. Treehugger Editors, "Why You Should Try Plastic-Free Dental Floss," Treehugger, August 29, 2022, https://www.treehugger.com/why-try-plastic-free-dental-floss-6501202. See also Ainsworth, Steve, "Flossed for Words: Developments in Oral Care," *Dental Nursing* 4, no. 12 (December 2008): 706–10, https://doi.org/10.12968/denn.2008.4.12.31784.

18. Cheprasov, Artem, "Are Floss Picks Bad for the Environment? Science Says Yes," Free RadiKal, April 26, 2022, https://www.freeradikal.com/blog/floss-picks-bad-environment/.

19. *National Geographic*, "How Your Toothbrush Became Part of the Plastic Crisis," accessed August 13, 2023, https://www.nationalgeographic.com/environment/article/story-of-plastic-toothbrushes.

20. Vadera, Shaili, and Soha Khan, "A Critical Analysis of the Rising Global Demand of Plastics and Its Adverse Impact on Environmental Sustainability," *Journal of Environmental Pollution and Management* 3, no. 1 (2021).

21. Shabecoff, Philip, "E.P.A. Sets Strategy to End 'Staggering' Garbage Crisis," *New York Times*, September 23, 1988, sec. U.S., https://www.nytimes.com/1988/09/23/us/epa-sets-strategy-to-end-staggering-garbage-crisis.html.

22. Rodriguez, Leah, "Which Period Products Are Best for the Environment?" Global Citizen, May 27, 2021, https://www.globalcitizen.org/en/content/best-period-products-for-the-environment/.

23. U.S. EPA, "WaterSense for Kids," February 21, 2023, https://www.epa.gov/watersense/watersense-kids.

24. BBVA, "Showering Daily," June 23, 2023, https://www.bbva.es/en/general/sostenibilidad/soluciones-para-personas/huella-de-carbono-personas/repositorio/ducharse-a-diario.html.

25. U.S. EPA, "WaterSense: Statistics and Facts," April 24, 2023, https://www.epa.gov/watersense/statistics-and-facts.

26. Geneva Environment Network, "Environmental Sustainability in the Fashion Industry," July 28, 2023, https://www.genevaenvironmentnetwork.org/resources/updates/sustainable-fashion/.

27. Geneva Environment Network, "Environmental Sustainability in the Fashion Industry."

28. Elliott, Debbie, "The 'Visible Mending' Trend of Fixing Clothes Can Be Traced to a Japanese Tradition," NPR, March 26, 2022, sec. Art & Design,

https://www.npr.org/2022/03/26/1088991078/the-visible-mending-trend-of-fixing-clothes-can-be-traced-to-a-japanese-traditio.

29. Richie, Marina, "Why Native Plants Are Better for Birds and People," Audubon, April 4, 2016, https://www.audubon.org/news/why-native-plants-are-better-birds-and-people.

30. Leahy, Stephen, "This Common Plastic Packaging Is a Recycling Nightmare," *National Geographic*, July 26, 2019, https://www.nationalgeographic.com/environment/article/story-of-plastic-common-clamshell-packaging-recycling-nightmare.

31. Pirog, Rich, Timothy Van Pelt, Kamyar Enshayan, and Ellen Cook, "Food, Fuel, and Freeways: An Iowa Perspective on How Far Food Travels, Fuel Usage, and Greenhouse Gas Emissions," Leopold Center for Sustainable Agriculture, Iowa State University, June 1, 2001, https://dr.lib.iastate.edu/entities/publication/51ef6421-3062-487f-af1a-29125328f7a5.

32. Mathers, Jason, "Green Freight Math: How to Calculate Emissions for a Truck Move," Environmental Defense Fund, March 24, 2015, https://business.edf.org/insights/green-freight-math-how-to-calculate-emissions-for-a-truck-move.

33. Global Citizen, "In 60 Seconds, 300 Football Fields Are Destroyed?" August 8, 2013, https://www.globalcitizen.org/en/content/in-60-seconds-300-football-fields-are-destroyed/.

BIBLIOGRAPHY

Ahmadi, Tala. "Replace Your Shampoo Bottle with a Shampoo Bar." The Boar, January 28, 2023. https://theboar.org/2023/01/replace-your-shampoo-bottle-with-a-shampoo-bar/#.

Ainsworth, Steve. "Flossed for Words: Developments in Oral Care." *Dental Nursing* 4, no. 12 (December 2008): 706–10. https://doi.org/10.12968/denn.2008.4.12.31784.

AppsRhino. "Food Delivery App | User Statistics 2022," March 29, 2023. https://www.appsrhino.com/blogs/food-delivery-app-user-statistics-2022.

Barone, Emily. "U.S. Plastic Recycling Rates Are Even Worse Than We Thought." *Time*, May 19, 2022. https://time.com/6178386/plastic-recycling-rates-overestimated/.

BBVA (Banco Bilbao Vizcaya Argentaria). "Showering Daily," June 23, 2023. https://www.bbva.es/en/general/sostenibilidad/soluciones-para-personas/huella-de-carbono-personas/repositorio/ducharse-a-diario.html.

Bennett, Sophia. "How to Recycle Ziploc Bags." Recycle Nation, October 7, 2014. https://recyclenation.com/2014/10/recycle-ziploc-bags/.

Birnbaum, Michael. "Plastic Forks Become Rarer as England Joins Global Effort to Ban Them." *Washington Post*, January 10, 2023. https://www.washingtonpost.com/climate-solutions/2023/01/10/england-single-use-cutlery-ban/.

Bloom, Richard, Marc Levine, and Robert Rivas. "School Facilities: Drinking Water: Water Bottle Filling Stations." Pub. L. No. AB 2638 (2022). California Legislative Information. https://leginfo.legislature.ca.gov/faces/billTextClient.xhtml?bill_id=202120220AB2638.

Booker, Linda, director. *STRAWS*. By the Brook Productions, 2017.

Bouyé, Mathilde, and David Waskow. "Climate Action Isn't Reaching the Most Vulnerable—But It Could." World Resources Institute, November 11, 2021. https://www.wri.org/insights/how-climate-action-can-help-vulnerable-populations.

Californians Against Waste. "AB 2026." Accessed June 1, 2023. https://www.cawrecycles.org/ab-2026.

CalRecycle. "Compost and Mulch Producers." Accessed May 26, 2023. https://calrecycle.ca.gov/organics/processors/.

Campos, Paulina. "9 Beauty Ingredients That Are Banned in Europe (But Legal in the U.S.)." *Byrdie*, January 28, 2022. https://www.byrdie.com/banned-ingredients-europe.

Carnegie, Dale. *How to Win Friends and Influence People.* Revised edition. New York: Pocket Books, 1982.

Center for Biological Diversity. "10 Facts About Single-Use Plastic Bags." Accessed June 22, 2023. https://www.biologicaldiversity.org/programs/population_and_sustainability/sustainability/plastic_bag_facts.html.

Centers for Disease Control and Prevention. "Increase Access to Drinking Water in Schools," April 16, 2021. https://www.cdc.gov/healthyschools/features/water_access.htm.

Chen, Grace. "How the Arts Benefit Your Children Academically and Behaviorally." Public School Review, February 13, 2023. https://www.publicschoolreview.com/blog/how-the-arts-benefit-your-children-academically-and-behaviorally.

Cheprasov, Artem. "Are Floss Picks Bad for the Environment? Science Says Yes." Free RadiKal, April 26, 2022. https://www.freeradikal.com/blog/floss-picks-bad-environment/.

Cheyenne Mountain Zoo. "What Happens to All Those Unwanted, Broken or Much Loved Crayons?" August 29, 2019. https://www.cmzoo.org/news/archive/what-happens-to-all-those-unwanted-broken-or-much-loved-crayons/.

Columbia University Irving Medical Center. "What Is a Plant-Based Diet, and Is It Healthy?" April 15, 2022. https://www.cuimc.columbia.edu/news/what-plant-based-diet-and-it-healthy.

Crayola. "How Many Crayola Products Do You Make per Year?" Accessed May 19, 2023. https://www.crayola.com/faq/your-business/how-many-crayola-products-do-you-make-per-year/.

Crayon Collection. "Crayon Recycling Program." Accessed May 22, 2023. https://crayoncollection.org/programs/crayon-recycling/.

Cuffe, Sandra. "2022 Amazon Fires Tightly Tied to Recent Deforestation, New Data Show." Mongabay, November 22, 2022. https://news.mongabay.com/2022/11/2022-amazon-fires-tightly-tied-to-recent-deforestation-new-data-show/.

Curry, David. "Uber Eats Revenue and Usage Statistics (2023)." Business of Apps, May 2, 2023. https://www.businessofapps.com/data/uber-eats-statistics/.

Davison, Courtney. "The Beginner's Guide to a Whole-Food, Plant-Based Diet." Forks Over Knives, July 28, 2023. https://www.forksoverknives.com/how-tos/plant-based-primer-beginners-guide-starting-plant-based-diet/.

Dean, Brian. "TikTok User Statistics (2023)." Backlinko, March 27, 2023. https://backlinko.com/tiktok-users.

Demeneix, Barbara A. "How Fossil Fuel-Derived Pesticides and Plastics Harm Health, Biodiversity, and the Climate." *The Lancet Diabetes & Endocrinology* 8, no. 6 (June 2020): 462–64. https://doi.org/10.1016/S2213-8587(20)30116-9.

Dykstra, Alan. "The Rhetoric of 'Whataboutism' in American Journalism and Political Identity." *Res Rhetorica* 7, no. 2 (2020). https://doi.org/10.29107/rr2020.2.1.

EARTHDAY.ORG. "Fact Sheet: Single-Use Plastics," March 29, 2022. https://www.earthday.org/fact-sheet-single-use-plastics/.

EARTHDAY.ORG. "Money in a Bottle," January 30, 2015. https://www.earthday.org/money-in-a-bottle/.

Elliott, Debbie. "The 'Visible Mending' Trend of Fixing Clothes Can Be Traced to a Japanese Tradition." NPR, March 26, 2022, sec. Art & Design. https://www.npr.org/2022/03/26/1088991078/the-visible-mending-trend-of-fixing-clothes-can-be-traced-to-a-japanese-traditio.

Environmental Defense Fund. "Overfishing: The Most Serious Threat to Our Oceans." Accessed June 21, 2023. https://www.edf.org/oceans/overfishing-most-serious-threat-our-oceans.

Fearing, Franklin. "Influence of the Movies on Attitudes and Behavior." *The ANNALS of the American Academy of Political and Social Science* 254, no. 1 (November 1947): 70–79. https://doi.org/10.1177/000271624725400112.

Fletcher, Angus. *Storythinking*. New York: Columbia University Press, 2023.

Fortune. "United Parcel Service." Accessed May 31, 2023. https://fortune.com/company/ups/.

Foss, Katherine A., and Brian G. Southwell. "Infant Feeding and the Media: The Relationship between *Parents' Magazine* Content and Breastfeeding, 1972–2000." *International Breastfeeding Journal* 1, no. 10 (2006): 10. https://doi.org/10.1186/1746-4358-1-10.

Foster, Joanna. "Farmers, Scientists Seek Solutions to Global Warming Caused by Cows." Environmental Defense Fund, April 15, 2022. https://vitalsigns.edf.org/story/farmers-scientists-seek-solutions-global-warming-caused-cows.

Gambino, Isabella, Francesco Bagordo, Tiziana Grassi, Alessandra Panico, and Antonella De Donno. "Occurrence of Microplastics in Tap and Bottled Water: Current Knowledge." *International Journal of Environmental Research and Public Health* 19, no. 9 (2022): 5283. https://doi.org/10.3390/ijerph19095283.

Gao, Angela L., and Yongshan Wan. "Life Cycle Assessment of Environmental Impact of Disposable Drinking Straws: A Trade-off Analysis with Marine Litter in the United States." *Science of The Total Environment* 817 (2022): 153016. https://doi.org/10.1016/j.scitotenv.2022.153016.

Gardiner, Beth. "The Plastics Pipeline: A Surge of New Production Is on the Way." *Yale Environment 360*, December 19, 2019. https://e360.yale.edu/features/the-plastics-pipeline-a-surge-of-new-production-is-on-the-way.

Gautam, Anirudh Muralidharan, and Nídia Caetano. "Study, Design and Analysis of Sustainable "Alternatives to Plastic Takeaway Cutlery and Crockery." *Energy Procedia* 136 (October 2017): 507–12. https://doi.org/10.1016/j.egypro.2017.10.273.

Generation Vegan. "Causes and Effects of Deforestation: How Can We Stop It?" GenV.org, February 9, 2021. https://genv.org/effects-of-deforestation/.

Geneva Environment Network. "Environmental Sustainability in the Fashion Industry," July 28, 2023. https://www.genevaenvironmentnetwork.org/resources/updates/sustainable-fashion/.

Global Citizen. "In 60 Seconds, 300 Football Fields Are Destroyed?" August 8, 2013. https://www.globalcitizen.org/en/content/in-60-seconds-300-football-fields-are-destroyed/.

Goddiess, Samantha. "The 10 Largest Toothpaste Brands in the United States." Zippia, April 25, 2023. https://www.zippia.com/advice/largest-toothpaste-brands/.

Golden, Jay S., Vairavan Subramanian, Gustavo Marco Antonio Ugarte Irizarri, Philip White, and Frank Meier. "Energy and Carbon Impact from Residential Laundry in the United States." *Journal of Integrative Environmental Sciences* 7, no. 1 (2010): 53–73. https://doi.org/10.1080/19438150903541873.

Goodchild, Anne, and Rishi Verma. "How Many Amazon Packages Get Delivered Each Year?" Supply Chain Transportation and Logistics Center at the University of Washington, October 17, 2022. https://depts.washington.edu/sctlctr/news-events/in-the-news/how-many-amazon-packages-get-delivered-each-year#.

Goodful. "Everybody's Dirty Little Secret." Facebook, November 20, 2020. https://www.facebook.com/watch/?v=156800712810150.

Grand View Research. "Plant-Based Meat Market Size, Share and Trends Analysis Report by Source (Soy, Pea, Wheat), by Product (Burgers, Sausages, Patties), by Type, by End-user, by Storage, by Region, and Segment Forecasts, 2023–2030," 2023. https://www.grandviewresearch.com/industry-analysis/plant-based-meat-market.

Groeger, Lena V. "Set It and Forget It: How Default Settings Rule the World." ProPublica, July 27, 2016. https://www.propublica.org/article/set-it-and-forget-it-how-default-settings-rule-the-world.

Habits of Waste. "#8meals." Accessed June 21, 2023. https://habitsofwaste.org/campaigns/8meals/.

Hellmann, Kai-Uwe, and Marius K. Luedicke. "The Throwaway Society: A Look in the Back Mirror." *Journal of Consumer Policy* 41, no. 1 (2018): 83–87. https://doi.org/10.1007/s10603-018-9371-6.

Holzer, Jillian. "Don't Put Vegetables in the Corner: Q&A with Behavioral Science Researcher Linda Bacon." World Resources Institute, June 12, 2017. https://www.wri.org/insights/dont-put-vegetables-corner-qa-behavioral-science-researcher-linda-bacon.

Howarth, Josh. "The Ultimate List of Beauty Industry Stats (2023)." Exploding Topics, March 23, 2023. https://explodingtopics.com/blog/beauty-industry-stats.

Igini, Martina. "The Environmental Impact of Online Shopping." Earth.Org, December 15, 2022. https://earth.org/online-shopping-and-its-environmental-impact/.

Jaglo, Kirsten, Shannon Kenny, and Jenny Stephenson. "From Farm to Kitchen: The Environmental Impacts of U.S. Food Waste." U.S. Environmental Protection Agency, November 2021. https://www.epa.gov/system/files/documents/2021-11/from-farm-to-kitchen-the-environmental-impacts-of-u.s.-food-waste_508-tagged.pdf.

Jamieson, Patrick E., and Daniel Romer. "Portrayal of Tobacco Use in Prime-Time TV Dramas: Trends and Associations with Adult Cigarette Consumption—USA, 1955–2010." *Tobacco Control* 24, no. 3 (2015): 243–48. http://www.jstor.org/stable/24842472.

Jarvie, Michelle. "Americans Waste Almost 50 Percent of Food Produced." Michigan State University Extension, November 21, 2014. https://www.canr.msu.edu/news/americans_waste_almost_50_percent_of_food_produced.

Karakartal, Demet. "Social Effects of Film and Television." *The Online Journal of New Horizons in Education* 11, no. 3 (2021). https://www.tojned.net/journals/tojned/articles/v11i03/v11i03-03.pdf.

Keoleian, Gregory, Martin Heller, and Diego Rose. "Implications of Future U.S. Diet Scenarios on Greenhouse Gas Emissions." Center for Sustainable Systems, University of Michigan, January 13, 2020. https://css.umich.edu/publications/research-publications/implications-future-us-diet-scenarios-greenhouse-gas-emissions.

Kim, Jinho, and Kerem Morgül. "Long-Term Consequences of Youth Volunteering: Voluntary Versus Involuntary Service." *Social Science Research* 67 (September 2017): 160–75. https://doi.org/10.1016/j.ssresearch.2017.05.002.

Langin, Katie. "Millions of Americans Drink Potentially Unsafe Tap Water. How Does Your County Stack Up?" Science, February 12, 2018. https://www.science.org/content/article/millions-americans-drink-potentially-unsafe-tap-water-how-does-your-county-stack.

Lazarus, David. "Column: You Do Know That, in Most Cases, Bottled Water Is Just Tap Water?" *Los Angeles Times*, September 28, 2021. https://www.latimes.com/business/story/2021-09-28/bottled-water-is-really-just-tap-water.

Leahy, Stephen. "This Common Plastic Packaging Is a Recycling Nightmare." *National Geographic*, July 26, 2019. https://www.nationalgeographic.com/environment/article/story-of-plastic-common-clamshell-packaging-recycling-nightmare.

Linnenkoper, Kirstin. "Colgate Leads Toothpaste Tube Recycling Innovation." Recycling International, June 28, 2019. https://recyclinginternational.com/plastics/colgate-leads-toothpaste-tube-recycling-innovation/26597/.

Litvinov, Amanda. "Out-of-Pocket Spending on School Supplies Adds to Strain on Educators." National Education Association, October 14, 2022. https://www.nea.org/advocating-for-change/new-from-nea/out-pocket-spending-school-supplies-adds-strain-educators.

Lonely Whale. "#StopSucking | Lonely Whale | For A Strawless Ocean." YouTube, 2017. https://www.youtube.com/watch?v=Q91-23B8yCg.

Los Angeles Department of Water and Power. "Plans to Install, Refurbish 200 Hydration Stations in Los Angeles Announced at 5th Annual Tap Water Day LA," June 13, 2019. https://www.ladwpnews.com/plans-to-install-refurbish-200-hydration-stations-in-los-angeles-announced-at-5th-annual-tap-water-day-la/.

Los Angeles Unified School District. "Los Angeles Unified Approves Infrastructure Upgrades and Sustainable Green Outdoor Learning Spaces," March 27, 2023. https://www.lausd.org/site/default.aspx?PageType=3&DomainID=4&ModuleInstanceID=4466&ViewID=6446EE88-D30C-497E-9316-3F8874B3E108&RenderLoc=0&FlexDataID=131211&PageID=1.

Mackintosh, Eliza. "The Amazon Is Burning Because the World Eats So Much Meat." CNN, August 23, 2019. https://www.cnn.com/2019/08/23/americas/brazil-beef-amazon-rainforest-fire-intl/index.html.

Mahoney, Adam. "In America's Cities, Inequality is Engrained in the Trees." Grist, May 5, 2021. https://grist.org/cities/tree-cover-race-class-segregation/.

Majd, Sanaz. "Should You Drink Tap or Bottled Water?" *Scientific American*, October 21, 2015. https://www.scientificamerican.com/article/should-you-drink-tap-or-bottled-water/.

Mandel, Kyla, and Brad Plumer. "One Thing You Can Do: Smarter Laundry." *New York Times*, October 2, 2019. https://www.nytimes.com/2019/10/02/climate/nyt-climate-newsletter-laundry.html.

MasterClass. "Film Career Guide: Above the Line vs. Below the Line Jobs," June 24, 2021. https://www.masterclass.com/articles/film-career-guide-above-the-line-vs-below-the-line-jobs#.

Mathers, Jason. "Green Freight Math: How to Calculate Emissions for a Truck Move." Environmental Defense Fund, March 24, 2015. https://business.edf.org/insights/green-freight-math-how-to-calculate-emissions-for-a-truck-move.

Mathieu, Edouard, and Hannah Ritchie. "What Share of People Say They Are Vegetarian, Vegan, or Flexitarian?" Our World in Data, May 13, 2022. https://ourworldindata.org/vegetarian-vegan.

McDonald, Robert I., Tanushree Biswas, Cedilla Sachar, Ian Housman, Timothy M. Boucher, Deborah Balk, David Nowak, Erica Spotswood, Charlotte K. Stanley, and Stefan Leyk. "The Tree Cover and Temperature Disparity in U.S. Urbanized Areas: Quantifying the Association with Income across 5,723 Communities." *PloS One* 16, no. 4 (2021): e0249715. https://doi.org/10.1371/journal.pone.0249715.

Miller, Chaz. "Plastic Film." Waste360, October 1, 2008. https://www.waste360.com/Recycling_And_Processing/plastic_film_ldpe.

Monitoring of the Andean Amazon Project. "MAAP #168: Amazon Fire Season 2022," November 3, 2022. https://maaproject.org/2022/amazon-fires-2022/.

Moraes, Marcela, John Gountas, Sandra Gountas, and Piyush Sharma. "Celebrity Influences on Consumer Decision Making: New Insights and Research Directions." *Journal of Marketing Management* 35, no. 13–14 (2019): 1159–92. https://doi.org/10.1080/0267257X.2019.1632373.

Moran, Alyssa J., Aviva Musicus, Mary T. Gorski Findling, Ian F. Brissette, Ann A. Lowenfels, S. V. Subramanian, and Christina A. Roberto. "Increases in Sugary Drink Marketing During Supplemental Nutrition Assistance Program Benefit Issuance in New York." *American Journal of Preventive Medicine* 55, no. 1 (July 2018): 55–62. https://doi.org/10.1016/j.amepre.2018.03.012.

Moreno, Gala D., Laura A. Schmidt, Lorrene D. Ritchie, Charles E. McCulloch, Michael D. Cabana, Claire D. Brindis, Lawrence W. Green, Emily A. Altman, and Anisha I. Patel. "A Cluster-Randomized Controlled Trial of an Elementary School Drinking Water Access and Promotion Intervention:

Rationale, Study Design, and Protocol." *Contemporary Clinical Trials* 101 (February 2021): 106255. https://doi.org/10.1016/j.cct.2020.106255.

Napurano, Clay. "California: Nation's First Statewide Plastic Bag Ban Cuts Waste." Frontier Group, August 27, 2021. https://frontiergroup.org/articles/california-nations-first-statewide-plastic-bag-ban-cuts-waste/.

NASA Global Climate Change: Vital Signs of the Planet. "Which Is a Bigger Methane Source: Cow Belching or Cow Flatulence?" Accessed June 21, 2023. https://climate.nasa.gov/faq/33/which-is-a-bigger-methane-source-cow-belching-or-cow-flatulence.

National Geographic. "How Your Toothbrush Became Part of the Plastic Crisis," accessed August 13, 2023. https://www.nationalgeographic.com/environment/article/story-of-plastic-toothbrushes.

National Oceanic and Atmospheric Administration (NOAA). "Sunscreen Chemicals and Marine Life." National Ocean Service, August 17, 2022. https://oceanservice.noaa.gov/news/sunscreen-corals.html.

National Resources Defense Council (NRDC). "The Truth About Tap," January 5, 2016. https://www.nrdc.org/stories/truth-about-tap.

Nepstad, Daniel C., Claudia M. Stickler, Britaldo Soares-Filho, and Frank Merry. "Interactions among Amazon Land Use, Forests and Climate: Prospects for a Near-Term Forest Tipping Point." *Philosophical Transactions of the Royal Society B: Biological Sciences* 363, no. 1498 (2008): 1737–46. https://doi.org/10.1098/rstb.2007.0036.

Newell, Jay, Charles T. Salmon, and Susan Chang. "The Hidden History of Product Placement." *Journal of Broadcasting & Electronic Media* 50, no. 4 (2006): 575–94. https://doi.org/10.1207/s15506878jobem5004_1.

Newstalk. "Over 56 Billion Coffee Capsules to Go to Landfill This Year," November 26, 2018. https://www.newstalk.com/news/over-56-billion-coffee-capsules-to-go-to-landfill-this-year-492445.

New York Times. "F.D.A Is Sued on Drug to Dry Mothers' Milk." August 17, 1994, sec. U.S. https://www.nytimes.com/1994/08/17/us/fda-is-sued-on-drug-to-dry-mothers-milk.html.

Oceana. "Amazon's Plastic Problem Revealed," December 15, 2020. https://doi.org/10.5281/ZENODO.4341751.

Oceana. "The Cost of Amazon's Plastic Denial on the World's Oceans," December 15, 2022. https://doi.org/10.5281/ZENODO.7525829.

Parker, Laura. "How the Plastic Bottle Went from Miracle Container to Hated Garbage." *National Geographic*, August 23, 2019. https://www.nationalgeographic.com/environment/article/plastic-bottles.

Pener, Degen. "Universal Filmed Entertainment Group Launches GreenerLight Sustainability Program (Exclusive)." *Hollywood Reporter*, March 9, 2023.

https://www.hollywoodreporter.com/business/business-news/universal-filmed-entertainment-greenerlight-sustainability-program-1235346745/.

Peters, Adele. "Can Online Retail Solve Its Packaging Problem?" Fast Company, April 20, 2018. https://www.fastcompany.com/40560641/can-online-retail-solve-its-packaging-problem.

Pirog, Rich, Timothy Van Pelt, Kamyar Enshayan, and Ellen Cook. "Food, Fuel, and Freeways: An Iowa Perspective on How Far Food Travels, Fuel Usage, and Greenhouse Gas Emissions." Leopold Center for Sustainable Agriculture, Iowa State University, June 1, 2001. https://dr.lib.iastate.edu/entities/publication/51ef6421-3062-487f-af1a-29125328f7a5.

Plastic Oceans International. "Plastic Pollution Facts." Accessed June 22, 2023. https://plasticoceans.org/the-facts/.

Poore, J., and T. Nemecek. "Reducing Food's Environmental Impacts through Producers and Consumers." *Science* 360, no. 6392 (June 1, 2018): 987–92. https://doi.org/10.1126/science.aaq0216.

Pruner, Aaron. "Above-the-Line vs. Below-the-Line Jobs in Film." Backstage, May 10, 2022. https://www.backstage.com/magazine/article/above-the-line-vs-below-the-line-crew-differences-74969/.

Quinton, Amy. "Cows and Climate Change." University of California, Davis, June 27, 2019. https://www.ucdavis.edu/food/news/making-cattle-more-sustainable.

Ramirez, Rachel. "Five Corporations Benefit from 25% of Bottled Water Sales." CNN, March 16, 2023. https://ix.cnn.io/charts/z6Wrz/2/.

Richie, Marina. "Why Native Plants Are Better for Birds and People." Audubon, April 4, 2016. https://www.audubon.org/news/why-native-plants-are-better-birds-and-people.

Rodriguez, Leah. "Which Period Products Are Best for the Environment?" Global Citizen, May 27, 2021. https://www.globalcitizen.org/en/content/best-period-products-for-the-environment/.

Rudd, Lauren F., Shorna Allred, Julius G. Bright Ross, Darragh Hare, Merlyn Nomusa Nkomo, Kartik Shanker, Tanesha Allen, et al. "Overcoming Racism in the Twin Spheres of Conservation Science and Practice." *Proceedings of the Royal Society B: Biological Sciences* 288, no. 1962 (2021): 20211871. https://doi.org/10.1098/rspb.2021.1871.

Rujnić Havstad, Maja, Ljerka Juroš, Zvonimir Katančić, and Ana Pilipović. "Influence of Home Composting on Tensile Properties of Commercial Biodegradable Plastic Films." *Polymers* 13, no. 16 (2021): 2785. https://doi.org/10.3390/polym13162785.

Same Side. Accessed May 30, 2023. https://onsameside.com/.

Schuman, Rebecca. "This Viral Formula Ad Absolves You for Using Formula." *Slate*, June 4, 2015. https://slate.com/human-interest/2015/06/a-century-of-formula-advertising-its-always-gone-straight-for-the-new-mother-jugular.html.

Sea Turtle Biologist. "Sea Turtle with Straw up Its Nostril: 'NO' TO SINGLE-USE PLASTIC." YouTube, 2015. https://www.youtube.com/watch?v=4wH878t78bw.

Shabecoff, Philip. "E.P.A. Sets Strategy to End 'Staggering' Garbage Crisis." *New York Times*, September 23, 1988, sec. U.S. https://www.nytimes.com/1988/09/23/us/epa-sets-strategy-to-end-staggering-garbage-crisis.html.

Shieber, Jonathan. "The #8meals App from Habits of Waste Helps People Cut Back on Meaty Meals to Save the Planet." TechCrunch, April 25, 2021. https://techcrunch.com/2021/04/25/the-8meals-app-from-habits-of-waste-helps-people-cut-back-on-meaty-meals-to-save-the-planet/.

Sony Pictures Entertainment. "Sony Pictures Entertainment to Eliminate Single-Use Plastic," August 3, 2020. https://www.sonypictures.com/corp/press_releases/2020/0803/sonypicturesentertainmenttoeliminatesingleuseplastic.

Spielberg, Steven, director. *E.T. the Extra-Terrestrial.* Universal Pictures, 1982.

Srinivasulu, M., M. Subhosh Chandra, Jaffer Mohiddin Gooty, and A. Madhavi. "Chapter 8—Personal Care Products—Fragrances, Cosmetics, and Sunscreens—in the Environment." In *Environmental Micropollutants*, edited by Muhammad Zaffar Hashmi, Shuhong Wang, and Zulkfil Ahmed, 131–49. Advances in Pollution Research. Amsterdam: Elsevier, 2022. https://doi.org/10.1016/B978-0-323-90555-8.00015-5.

Statista. "Advertising Worldwide—Statistics and Facts." Accessed June 22, 2023. https://www.statista.com/topics/990/global-advertising-market/.

Statista. "Global: Traditional TV and Home Video Number of Users 2018–2027," June 1, 2023. https://www.statista.com/forecasts/1207931/tv-viewers-worldwide-number.

Stockton Recycles. "Plastic Utensils Go in the Trash," March 5, 2020. https://stocktonrecycles.com/plastic-utensils-go-in-the-trash/.

Surfrider Foundation. "City of Los Angeles Plastic Bag Ban," June 6, 2013. https://www.surfrider.org/campaigns/City%20of%20Los%20Angeles%20Plastic%20Bag%20Ban.

Tamer, Mary. "On the Chopping Block, Again." *Harvard Ed. Magazine*, June 8, 2009. https://www.gse.harvard.edu/news/ed/09/06/chopping-block-again.

Thompson, Don. "Think Those Bags Are Recyclable? Think Again." Associated Press, December 29, 2022. https://apnews.com/article/california-state-government-business-climate-and-environment-59dc96edb1b2901f829071a6f9153df1.

Triantafillopoulos, Nick, and Alexander Koukoulas. "The Future of Single-Use Paper Coffee Cups: Current Progress and Outlook." *BioRes* 15, no. 3 (2020). https://bioresources.cnr.ncsu.edu/resources/the-future-of-single-use-paper-coffee-cups-current-progress-and-outlook/.

Tuman, Nina. "Why You Should Switch to Reusable Cleaning Bottles." Tru Earth. Accessed June 24, 2023. https://www.tru.earth/Why-You-Should-Switch-to-Reusable-Cleaning-Bottles.

U.S. Bureau of Labor Statistics. "Consumer Expenditures in 2020," BLS Report 1096, December 2021. https://www.bls.gov/opub/reports/consumer-expenditures/2020/home.htm.

U.S. Department of Education. Title I, Part A Program, November 7, 2018. https://www2.ed.gov/programs/titleiparta/index.html.

U.S. Department of Energy. "LED Lighting." Accessed August 3, 2023. https://www.energy.gov/energysaver/led-lighting.

U.S. Environmental Protection Agency (EPA). "EPA Report Shows Disproportionate Impacts of Climate Change on Socially Vulnerable Populations in the United States." September 2, 2021. https://www.epa.gov/newsreleases/epa-report-shows-disproportionate-impacts-climate-change-socially-vulnerable.

U.S. EPA. "National Overview: Facts and Figures on Materials, Wastes and Recycling." Overviews and Factsheets, October 2, 2017. https://www.epa.gov/facts-and-figures-about-materials-waste-and-recycling/national-overview-facts-and-figures-materials.

U.S. EPA. "Plastics: Material-Specific Data." Collections and Lists, September 12, 2017. https://www.epa.gov/facts-and-figures-about-materials-waste-and-recycling/plastics-material-specific-data.

U.S. EPA. "Tailpipe Greenhouse Gas Emissions from a Typical Passenger Vehicle," January 12, 2016. https://www.epa.gov/greenvehicles/tailpipe-greenhouse-gas-emissions-typical-passenger-vehicle.

U.S. EPA. "WaterSense for Kids," February 21, 2023. https://www.epa.gov/watersense/watersense-kids.

U.S. EPA. "WaterSense: Statistics and Facts," April 24, 2023.

Vadera, Shaili, and Soha Khan. "A Critical Analysis of the Rising Global Demand of Plastics and Its Adverse Impact on Environmental Sustainability." *Journal of Environmental Pollution and Management* 3, no. 1 (2021).

Vargas, Annette M. "Arts Education Funding." *Journal of Women in Educational Leadership* (2017). https://doi.org/10.13014/K21Z42K0.

Veganuary. Accessed August 3, 2023. https://veganuary.com/en-us/.

Wansink, Brian, and Koert van Ittersum. "Portion Size Me: Downsizing Our Consumption Norms." *Journal of the American Dietetic Association* 107, no. 7 (2007). https://papers.ssrn.com/abstract=2474331.

Wansink, Brian, Koert van Ittersum, and James E. Painter. "Ice Cream Illusions: Bowls, Spoons, and Self-Served Portion Sizes." *American Journal of Preventive Medicine* 31, no. 3 (2006): 240–43. https://doi.org/10.1016/j.amepre.2006.04.003.

World Wildlife Fund (WWF), "Unsustainable Cattle Ranching." Accessed August 3, 2023. https://wwf.panda.org/discover/knowledge_hub/where_we_work/amazon/amazon_threats/unsustainable_cattle_ranching/.

WWF. "Will There Be More Plastic Than Fish in the Sea?" Accessed June 22, 2023. https://www.wwf.org.uk/myfootprint/challenges/will-there-be-more-plastic-fish-sea.

INDEX

Page references for figures are italicized.

ABOUT THE AUTHOR

Sheila Michail Morovati was born in Tehran, Iran, and immigrated to the United States after the 1979 Islamic Revolution. She is the founder and president of two environmentally focused nonprofit organizations, Crayon Collection and Habits of Waste (HOW). With a sociology degree from UCLA, she has been successfully growing her nonprofits since 2013 by creating systems changes so that the masses can partake in environmentalism. She is passionate about solving the climate crisis by promoting "imperfect environmentalism" as an avenue to increase involvement from everyday individuals who feel climate anxiety yet feel powerless to make change.

Her first nonprofit, Crayon Collection, has collected over twenty million gently used crayons from restaurant chains worldwide, which were then donated to vulnerable schools worldwide. The organization set a Guinness World Record in August 2018 for the largest crayon donation in history with over one million crayons donated to over nine hundred teachers in Los Angeles. Habits of Waste, Sheila's second non-profit, is focused on helping everyday individuals become "imperfect environmentalists" through her many campaigns. Sheila spearheaded the first ban of plastic straws, utensils, and stirrers in the world in the City of Malibu and has been credited with leading the movement against single-use plastic straws and cutlery. Next, through her #CutOutCutlery campaign, she convinced Uber Eats, Postmates, Grubhub, and DoorDash to change their default settings globally so that users receive plastic cutlery upon request only. This campaign has prevented billions of packs of plastic cutlery from entering the waste stream, saved restaurants millions of dollars in expenses, and led to legislation in California (AB 1276), Washington, and New York requiring single-use foodware items to be provided upon request only. Sheila has also spearheaded several other legislative efforts, including California (AB 2638), which requires schools throughout the state of California to provide a minimum of one water bottle filling station for each school undergoing modernization, and a minimum of one water bottle filling station per 350 people at each new school being constructed. This developed after Sheila unveiled the dilapidated water fountains inside public schools, which inevitably forced students to depend on costly plastic water bottles for rehydration.

Hollywood is also creating change thanks to Sheila's efforts as the leader in removing all single-use plastic from being shown on television and in film. Sheila created the #LightsCameraPlastic? campaign to denormalize single-use plastic and believes it will have an effect similar to the decline in the number of smokers after cigarettes were removed from Hollywood productions. Major motion pictures (e.g., *Marry Me* starring Jennifer Lopez) and major studios have adopted this campaign to continue removing all plastics from television and film sets.

Upon learning that adopting plant-based diet is the single most important commitment individuals can make to combat climate change, Sheila created an application that helps individuals be successful at eating a partially plant-based diet. She translated a published study from the University of Michigan into something the masses can partake in. Sheila also created the partially plant-based campaign called #8meals and developed the associated app as a free support tool that shares the carbon offset of the effort. Sheila is regularly invited to participate in the World Economic Forum's Nutrition Disruptors and Consumers for Climate Action, as well as recently joining a smaller group of environmental leaders in the Global Future Council on Net Zero Living, which was cochaired by the UCLA chair of sociology, Dr. Ed Walker, and the chief sustainability officer (CSO) of Google, Kate Brandt.

Sheila lives in Los Angeles with her husband and two children and has been featured in *People* magazine, the *Los Angeles Times*, the *Holly-wood Reporter*, *Variety*, BuzzFeed, ATTN: media, *Forbes*, *Huffington Post*, *Bloomberg Radio*, and various regional broadcasts. She has appeared on national television shows such as the *Kelly Clarkson Show* and *CBS Sunday Morning*.

Interested in learning more about becoming an imperfect environmentalist? Visit www.imperfectenvironmentalist.com for more information and tools.